国家出版基金项目

『十二五』国家重点图书出版规划项目

中国古建筑测绘大系·陵寝建筑与祠庙建筑

关外三陵和北镇庙

哈尔滨工业大学建筑学院 编写

侯幼彬 刘大平 刘洋 主编

Traditional Chinese Architecture Surveying and
Mapping Series:
Mausoleum Architecture & Shrines and Temples Architecture

THREE MAUSOLEUMS OUTSIDE
SHANHAIGUAN PASS AND BEIZHEN TEMPLE

Compiled by School of Architecture, Harbin Institute of Technology
Edited by HOU Youbin, LIU Daping, LIU Yang

China Architecture & Building Press

中国建筑工业出版社

Contents

目 录

Introduction

导 言

Among the three mausoleums outside Shanhaiguan Pass, Yong mausoleum is the most different one. It was used to be an ancestral grave of the local officials at the beginning of its construction. After Kangxi period, it was extended continually, so the pattern of Yong mausoleum formed finally. From the building's characteristics of Fu mausoleum and Zhao mausoleum, we could find the way of which Man nationality and Han nationality were combined by royal family of Qing dynasty.

There are constitution factors of mausoleum of Ming dynasty in Fu mausoleum and Zhao mausoleum. Which includes Longen gate, Longen hall, stone arch way, the main red gate, stone statues, stove for burning, Minglou, stele pavilion, two-column archway and stone altar for offering.

The layout pattern of Fu mausoleum and Zhao mausoleum is to develop in depth along the axial line which has adopted by mausoleum of Ming dynasty. The Leading space in front of god path is open and sparse. The rear area is ritualistic space which is introverted and assembled by buildings and countyards.

The relationship between these two mausoleums and environment, as the mausoleum of Ming dynasty, follows the principles of geomantic omen theory. Artificial environment and natural environment complement each other.

There are also some differences in composition between the mausoleums of Ming dynasty and these two mausoleums. Fu mausoleum and Zhao mausoleum are far from each other, so that the mausoleum area is not formed. The main buildings show the features of castle style. Tower style is more popular, Longen gate as the main gate of Fangcheng is a tower with

关外三陵中，永陵情况特殊，初建时是作为地方官的祖坟而非皇族祖陵的，直到康熙以后不断增建才具备了陵的格局。福陵和昭陵的建筑特色可以较多地反映出清朝皇族在当时特定的历史时期将满、汉文化杂糅的做法。

福、昭二陵具备明陵的主要构成要素，包括隆恩门、隆恩殿、石牌坊、正红门、石象生、焚帛炉、明楼、碑亭、二柱门、石五供等。

在总体布局上，福、昭二陵沿袭了明陵的轴线纵深布局，前区的神道形成疏朗、开阔的前导性空间，后区用建筑与庭院围合成集聚、内向的仪式性空间。

在环境处理方面，福、昭二陵与明陵一样遵循着传统陵墓的风水原则，体现了人工环境与自然环境的相得益彰。

与明陵的构成也有若干的不同之处，如福、昭二陵相隔很远，没有形成陵

three storeys, the four corners of Fangcheng is occupied by turrets of two layers too, and there also are a couple of side towers in the side hall of Zhao mausoleum. The pattern of the main hall and wing-houses is always used in style of one main and four supporting buildings.

The construction method of these two mausoleums is also different from mausoleums inside Shanhaiguan pass. Such as the beam frame constituted by Lin-jiu purlin, diversified Dougong, glaze component with abundant rural details, the Xumizuo of unofficial form and so on. These styles which stand for local customs is a kind of art failure, but unique cultural value could be also refined from them.

区；主体建筑采用城堡式格局；崇尚楼阁，方城正门隆恩门均为三层城门楼，方城的四角用两层的角楼，甚至昭陵的配殿也设有一对楼阁；主体建筑与配殿多采用『一正四厢』的格局。

在构造做法上，福、昭二陵也有一些不同于关内陵的做法，如采用檩枋组合的梁架，多样化的斗栱形制，带有丰富乡土细部的琉璃构件，非定型式的须弥座，等等。这些做法带有俚俗的格调并夹杂艺术的败笔，但有着独特的文化价值。

图

版

Figure

清永陵

Yong Mausoleum

项目信息

地　　址　辽宁省抚顺市新宾满族自治县城西郊

始建年代　1589年（明万历二十六年）

占地面积　1.1万平方米

主管单位　新宾满族自治县文化广电局

测绘单位　哈尔滨建筑工程学院（现哈尔滨工业大学）

测绘时间　1981年

Project Information

Location: Western suburb, Xinbin Manchu Autonomous county, Fushun prefecture, Liaoning province

Construction Date: 1589, the twenty six year of Wanli reign period of the Ming dynasty

Area: 11,000 square meters

Administrative Office: Xinbin Manchu Autonomous County Culture and Broadcasting and Television Bureau

Responsible Department: Harbin Institute of Architecture and Engineering (Harbin Institute of Technology)

Survey Time: 1981

1. Brief History

Yong Mausoleum of Qing Dynasty is an ancestor Mausoleum for Nurhachi family which was first built in the 26th year of Wanli in Ming Dynasty (1598). In 1634, emperor Huang Taiji (titled Taizong of Qing) decreed the Fuman family tomb as Xingjing Mausoleum. In the 5th year of Shunzhi (1648), emperor Fulin (titled Shizu of Qing) conferred Möngke Temür as the "Yuan emperor of Zhaozu", Fuman as the "Zhi emperor of Xingzu", Juechang'an as the "Yi emperor of Jingzu" and Takeshi as the "Xuan emperor of Xianzu". The sacrificial hall, the side hall and the rampart of Fangcheng were built in the 10th year of Shunzhi (1653) and the Shengong-shengde stele and the pavilion were set up for the ancestor Zhao and Xing in 1655. The ancestor tombs of Jing and Xian emperor, together with the tombs of Juechang'an's sons Lidun and Tacha, were moved to the Xingjing Mausoleum of ancestor Zhao and Xing, hence Yong Mausoleum is called "Mausoleum of Four Ancestors". In the 16th year of Shunzhi (1659) the Xingjing Mausoleum were renamed Yongling. In 1661, the sacrificial hall was named "Qiyun" Hall and the gate of Fangcheng named "Qiyun" Gate, and built Shengong-shengde stele and pavilion for the ancestors Jing and Xian. Yellow glazed tiles replaced grey tiles of Yong Mausoleum in the 16th year of Kangxi (1677). The Qiban (guards' house) and Zhuban were built in the 8th year of Yongzheng (1730), while the kitchen and washhouse were built in the 1st year of Qianlong (1736).

2. General Layout

Yong Mausoleum is located at Yong Mausoleum town, Xinbin county, Liaoning province, where Qiyun Mountain is to the north, Suzi River to the south and Yancong Mountain across the river (Fig.1). The Mausoleum consists three parts of front court, central court and Baocheng, and is surrounded by red wall. Beyond the red wall there are serving and guarding facilities as the pavilion for butchering sacrificial animals, dismounting stele and boundary stones colored in red, white and blue. Yong Mausoleum adopts the smallest scale but largest land area among the "Three Mausoleums outside Shanhaiguan Pass".

The center of the south of the front court is the front red gate, and after the gate Qiban (guards' house), Zhuban, kitchen and washhouse are arranged every two on each side of the path. Four stele pavilions housing tablets stand in line facing to south in the center with Qiyun Gate in the north of the front court.

一、历史沿革

清永陵是努尔哈赤家族的祖陵，始建于明万历二十六年（1598年），明崇祯七年（1634年）清太宗皇太极敕福满族墓称兴京陵。顺治五年（1648年）清世祖福临追封猛哥贴木尔为"肇祖原皇帝"，福满为"兴祖直皇帝"，觉昌安为"景祖翼皇帝"，塔克世为"显祖宣皇帝"。顺治十二年（1655年）立肇、兴二祖神功圣德碑、建碑亭。顺治十五年（1658年）始建享殿、配殿、方城门墙。顺治十六年（1659年）将景、显二祖陵及礼敦、塔察二墓迁至兴京陵肇、兴二祖墓前，永陵自此成为"四祖陵"。顺治十六年（1659年）兴京陵更名为永陵。顺治十八年（1661年）命名享殿为"启运殿"，方城门为"启运门"，立景、显二祖神功圣德碑、建碑亭。康熙十六年（1677年）永陵改青瓦为黄色琉璃瓦。雍正八年（1730年）建齐班房、祝版房。乾隆元年（1736年）建茶膳房、涤器房。

二、总体布局

永陵位于辽宁新宾县永陵镇，北依启运山，南临苏子河，与烟筒山隔河相望（图1）。陵园由前院、

图2 《亚细亚大观》载永陵碑亭

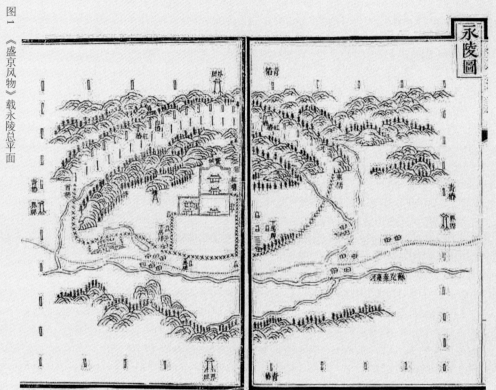

图一 《盛京风物》载永陵总平面

Fig.1 The master plan of Yong mausoleum recorded in *The scenery of Shengjing*
Fig.2 The stele pavilions of Yong mausoleum recorded in *The great view of Asia*

Behind Qiyun Gate is the main court of Yong Mausoleum, known as "Fangcheng". The Qiyun Hall is centered in north and side halls in east and west. There is a furnace at the west front of Qiyun Hall platform, and a smaller size Baocheng at the back of Qiyun Hall.

3. Buildings

The front red gate, known as palace gate, with three bays in width and two bays in depth, is a central-columned gate. The low stylobate is attached by the front and back steps with dropping belts. Double-leafs doors are set between the central columns of three bays and the beam frame is Lin-jiu Purlin. Chitou, the porch heads, extend from the front and back gable, with Pantou embedded by glazed decorative tiles supporting the front eave. Yellow glazed tiles cover the flush gable roof, with the main ridge decorated by dragons playing with a ball, and the dropping ridges decorated by one immortal figure and two animal figures.

Four ancestors' stele pavilions housing tablets are arranged in the position order of Zhao, Xing, Jing and Xian, from old to young, center to side and left to right. The four stele pavilions adopt the same size, square plan and three bays both in width and depth. Brick wall is constructed at the west, and arches are in the center of south and north. The low stylobate is attached by the front and back steps with dropping belts. At the corner under gable inlayed stone slab and corner pier caved with walking dragon, rising dragon and fallen dragon. Three intermediate bracket sets with seven-tier parallel arms are used in the central bay while no intermediate sets in the side bays. The gable-and-hip roof is decorated with curly-tailed hollow ornaments at the ends of the main ridge, and animal figures with Xiezhi's head at the ends of vertical and diagonal ridges (Fig.2).

Qiyun Gate is a gate hall with a row of central columns as well as surrounding colonnade, which is three bays in width and two bays in depth. The low stylobate is attached by the front and back steps using three connected stairs with imperial path. Two groups of structural frames in central bay are of seven-purlins system with central columns and front and back colonnade, and the structural frames at the gables add two more columns thus forming seven columns standing on the stylobate. The method of purlin-plate-lintel and Lin-jiu purlins are adopted mixedly, without bracket set (Fig.3). Three double-leaf doors are placed between the central columns, with solid wooden leafs painted red but without gap boards either on right or left. Walking dragon and cloud pearl are decorated on the main

三、单体建筑

正红门俗称『宫门』，面阔三间，进深二间，是中柱无廊的载门。台基低矮，前后正阶出踏跺。三开间的中柱之间都设双扇栅门，梁架为檩枋做法。山墙前后出墀头，前檐盘头上嵌琉璃花饰馋檐砖。采用黄色琉璃瓦硬山顶，正脊饰游龙戏珠，垂脊兽前用一仙人、两走兽。

四祖碑亭按中长次幼、左老右少的位序依次为肇、兴、景、显四祖。四亭大小形制相同，正方形平面，面阔、进深均为三间。西面包砌砖墙，南北正中均辟券门。石台基低矮，前后出垂带踏跺。明间用三攒平身科斗口三翘七踩品字斗栱，次间无平身科。屋顶为单檐歇山式，透空卷尾正吻，有獬首形的垂、馋兽（图 2）。

启运门内为永陵主院，俗称『方城』。启运殿居中靠北，左右设东西配殿。启运殿月台前西侧有一座焚帛炉。启运殿后部即小尺度的宝城。

前院南面正中为正红门，门内神道两侧按『一正四厢』格局分别布置有齐班、祝版房和茶膳、涤器房，院中心坐北朝南一字排开四座碑亭，前院正北为启运门。

青三色界碑等。在『关外三陵』中，永陵陵园规模最小，但陵区占地最广。

中院、宝城三部分组成，四周有红墙围护。红墙外围的服务和防护设施有宰牲亭、下马碑及红、白、

图 3 《亚细亚大观》载永陵启运门

图 4 《亚细亚大观》载永陵启运门袖壁

图 5 《亚细亚大观》载永陵东配殿

Fig.3 The Qiyun gate of Yong mausoleum recorded *in The great view of Asia*
Fig.4 The sleeve wall besides Qiyun gate of Yong mausoleum recorded in *The great view of Asia*
Fig.5 The east side hall of Yong mausoleum recorded in *The great view of Asia*

ridge of the glazed-tiled gable-and-hip roof. At the ends of the main ridge, the sword handles of the curly-tailed hollow ornaments are shaped as flame circle, and inlayed two Chinese characters "sun" and "moon" in the voids. One immortal and five animal figures are placed in front of the "Qiangshou"(big animal figures) of the diagonal ridge. Sleeve walls extend on both left and right sides of the court wall, whose lower part is brick-built Sumeru stylobate. Glazed-tiled overhanging gable roof is crowned on the top, with colored glazed curled-up dragon inlayed in the central Begonia panel, and glazed cloud, flower and plant patterns inlayed in the four corners (Fig.4).

Both east and west side halls are the same scale of three bays in width, with gable–and-hip roof, and surrounding colonnade. The stylobate of side hall is low and the front steps are with dropping belts. Two groups of structural frames in central bay are of seven-purlins system with front and back colonnade, and the structural frames at the gables have seven columns standing on the stylobate. There places a vase-shaped supporting component on the 5-purlin beam, and the method of purlin-plate-lintel and Lin-jiu purlins are adopted mixedly, without bracket set. In the front, between columns of the second row, partition leafs are set in the central bay and sill wall and sill windows are used in side bays. Inside the hall all structural components are exposed without ceiling (Fig.5).

The Qiyun Hall is the Sacrificial hall of Yong Mausoleum, with three bays both in width and depth, gable–and-hip roof, and surrounding colonnade, standing on a stylobate with a large platform enclosing from three sides of it. Three stairs are set in front of the platform and the central one is with imperial path without decoration (Fig.6). 7-purlin beam is adopted in two groups of structural frames of the central bay. Under the 7-purlin beam, two reinforce columns are added near the back inner column and an octagon-section reinforce column are added close to the front inner column. In the colonnade the tie-beam is appressed the bottom of Baotou beam (short beam) without bracket set. In the front, between columns of the second row, partition leafs are set in the central bay and sill wall and sill windows are used in side bays. In the back, double-leaf grid-framed door is in the middle of the central bay. Inside the hall, all structural components are exposed without ceiling, and there are four big and small warm rooms, as well as the throne with dragon and phoenix patterns, five-offering table and lanterns. Xuanzi color paintings are adopted for both interior and exterior of the hall.

Baocheng locates behind Qiyun Hall, with a horseshoe-shaped plan, and a "sacred tree" in the front and octagonal arc wall at back. It is divided into two layers that are connected

启运门是面阔三间、进深两间、外带周围廊的中柱式门殿。台基低矮，前后用三出陛连三带御路踏跺。明间两缝梁架用七檩中柱前后廊构架，山金缝梁架在中柱前后增设了两金柱形成七柱落地，檩垫枋做法与檩枋做法混用，无斗栱（图3）。顺中柱开三间大门，每间安双扇朱漆实榻门，无左右余塞板。单檐歇山式琉璃瓦顶，正脊饰行龙、云珠，透空卷尾正吻，每间上砌圆形火焰圈，圈内分别镶『日』『月』二字，饕餮兽前用一仙人、五走兽。左右两侧院墙各出一道袖壁，袖壁下砌砖须弥座，上冠悬山式琉璃瓦顶，中心海棠盒子镶嵌五彩琉璃蟠龙，四岔角镶琉璃朵云及花草（图4）。

东、西配殿为形制相同的面阔三间、带周围廊的单檐歇山顶建筑。配殿台基低矮，正阶出垂带踏跺。明间两缝梁架用七檩前后廊，山金缝梁架则七柱落地，五架梁中部加垫宝瓶，檩垫枋做法与檩枋做法混用，无斗栱。前檐做金里装修，明间用槅扇，次间用槛墙槛窗，殿内为彻上明造（图5）。

启运殿为永陵的享殿，面阔、进深各三间，带周围廊，单檐歇山顶。下出一层台基和宽大的月台，月台三面环包台基，月台前出三出陛，中为御路踏跺，御路石素面无饰（图6）。明间两缝梁架用七架梁下后上金柱位置增加了两根加固柱，贴前金柱内侧增加了八角形断面加固柱，廊架梁做法，七架梁下后上金柱位置增加了两根加固柱，贴前金柱内侧增加了八角形断面加固柱，廊

图 6 《亚细亚大观》载永陵启运殿

Fig.6 The Qiyun hall of Yong mausoleum recorded in *The great view of Asia*

by the imperial path and steps at the front. On the upper layer, there buried ancestor Xing of Fuman in the middle, Jing of Juechang'an on the left, and Xian of Takeshi on the right. The cenotaph of Möngke Temür locates to the northeast of the Xing Mausoleum. On the lower layer, there buried "Wugong Prince" Lidun on the left and "Kegong Prince" Tachapiangu on the right. "Qiyun Mountain", the mountain of Mausoleum is just behind Baocheng.

参考文献
References

[1] 王伯扬. 中国古建筑大系（第 2 册）：帝王陵寝建筑 [M]. 北京：中国建筑工业出版社，1993.

[2] 潘谷西. 中国古代建筑史（第四卷）：元明建筑（第二版）[M]. 北京：中国建筑工业出版社，2009.

[3] 陈伯超，刘大平，李之吉. 中国古代建筑史全集（辽黑吉卷）[M]. 北京：中国建筑工业出版社，2016.

[4] 村田治郎. 亚细亚大观 [J]. 亚细亚写真大观，1942.

[5] 侯幼彬. 读建筑 [M]. 北京：中国建筑工业出版社，2012.

[6] 辽宁省图书馆. 盛京风物 [M]. 北京：中国人民大学出版社，2006.

[7] 黑龙江省博物馆. 中东铁路大画册 [M]. 哈尔滨：黑龙江人民出版社，2013.

山『启运山』。

内随梁枋紧贴抱头梁下皮，无斗栱。前檐做金里装修，明间用槅扇，次间用槛墙槛窗，后檐明间正中设双扇对开的棋盘门。殿身内彻上明造，殿内有大、小暖阁各四座，另有龙凤宝座、五供案桌和朝灯等陈设。殿内外绘旋子彩画。

启运殿后即为宝城。宝城平面呈马蹄形，前部有『神树』，后用八角弧形罗圈墙。分上、下两层台地，用正阶御路踏跺相接。上层台中葬兴祖福满、左昭景祖觉昌安、右穆显祖塔克世。肇祖孟特穆衣冠冢位于兴祖墓东北。下层台左葬武功郡王礼敦，台右葬恪恭贝勒塔察篇古。宝城后即陵

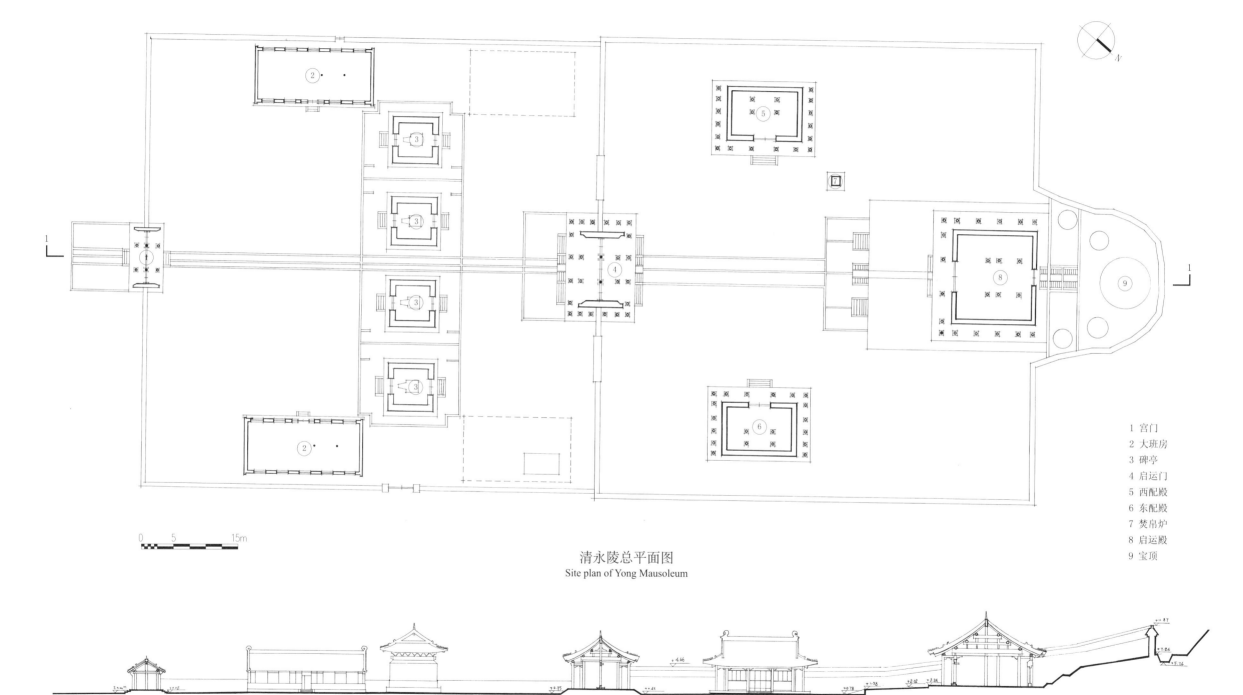

1 宫门
2 大班房
3 碑亭
4 启运门
5 西配殿
6 东配殿
7 焚帛炉
8 启运殿
9 宝顶

0　5　15m

清永陵总平面图
Site plan of Yong Mausoleum

0　5　15m

清永陵 1-1 剖面图
1-1 section of Yong Mausoleum

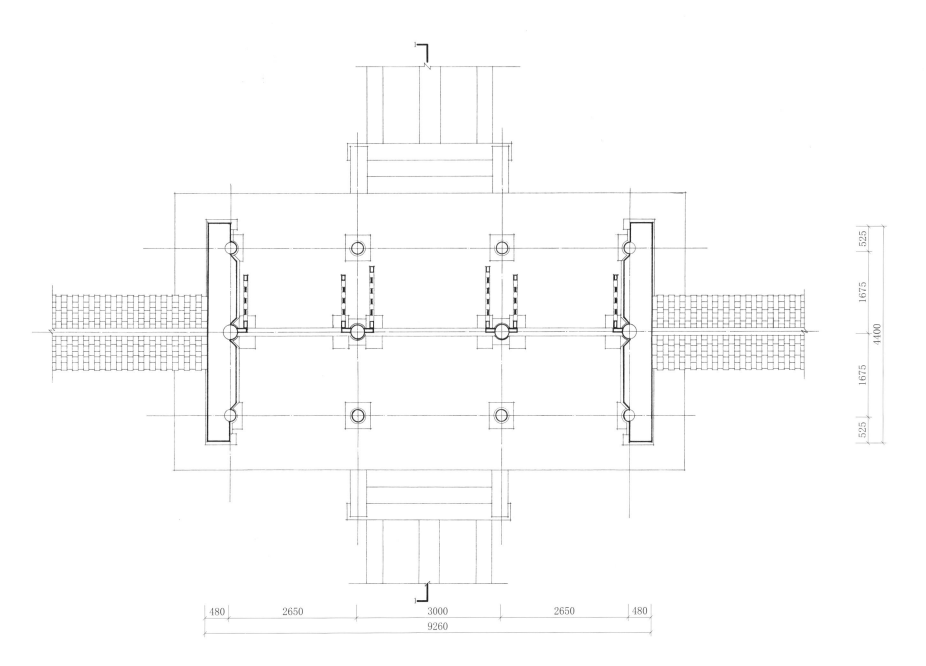

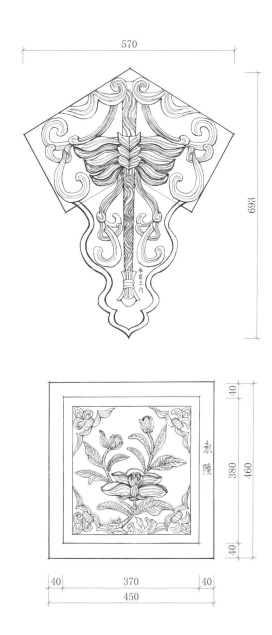

清永陵宫门平面图
Plan of palace gate in Yong Mausoleum

清永陵宫门琉璃大样图
Detail drawing of glaze component in Yong mausoleum

0   0.5   3m

清永陵宫门正立面图
Elevation of palace gate in Yong Mausoleum

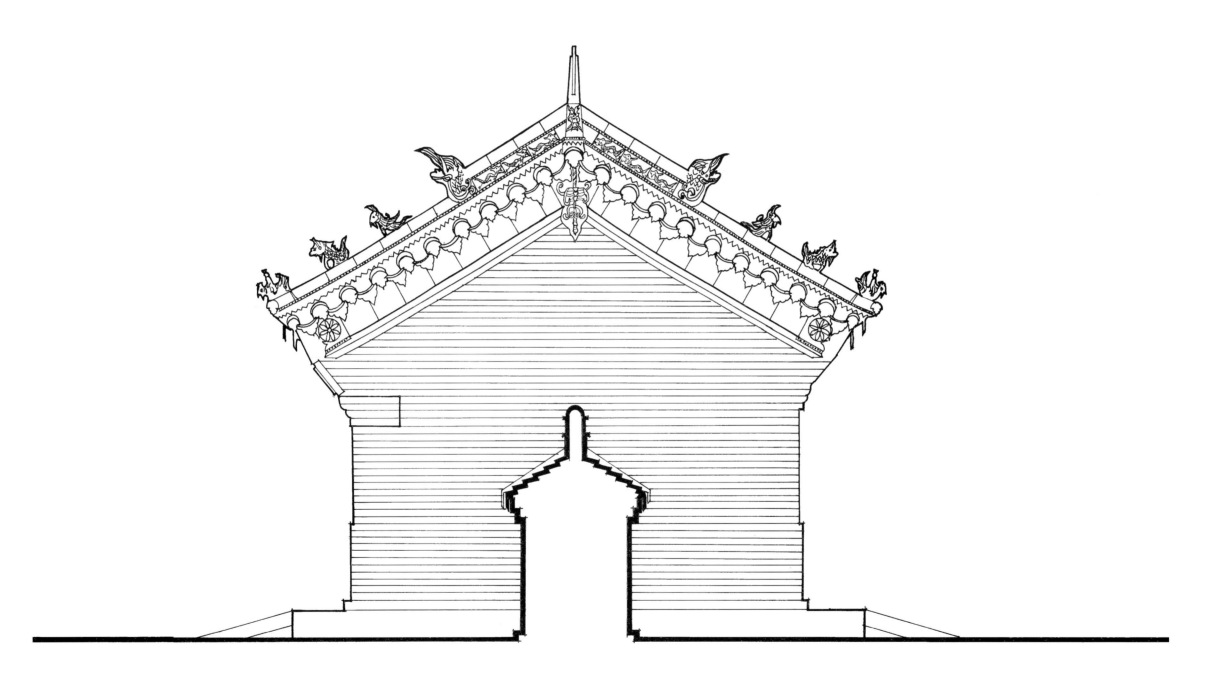

清永陵宫门侧立面图
Side elevation of palace gate in Yong Mausoleum

0   0.5   3m

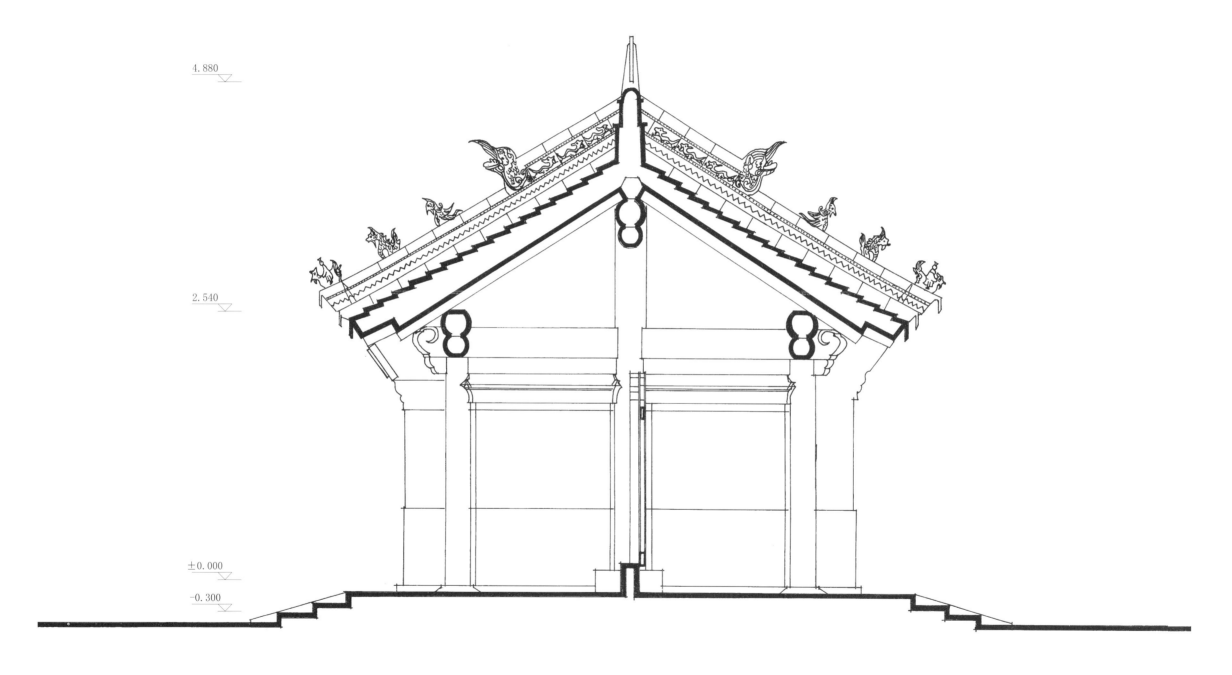

4.880

2.540

±0.000

-0.300

0   0.5   3m

清永陵宫门 1-1 剖面图

1-1 section of palace gate in Yong Mausoleum

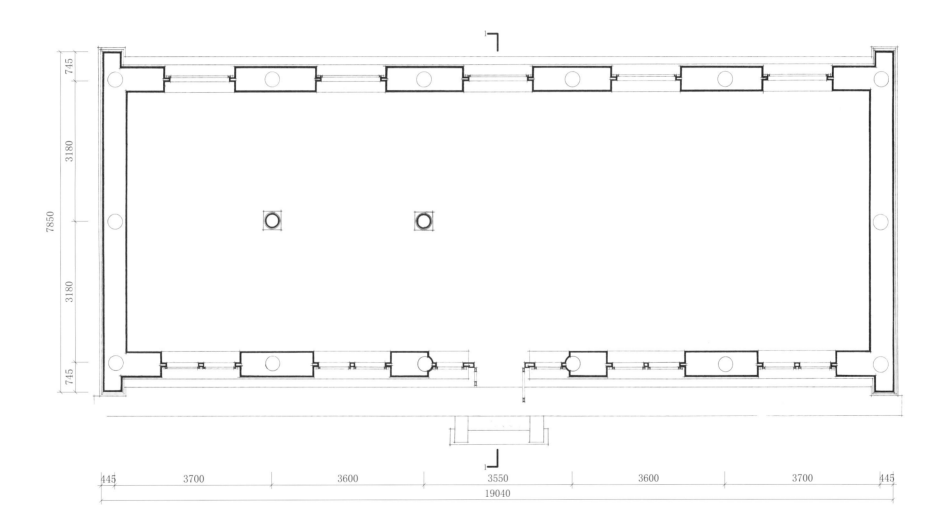

清永陵大班房平面图

Plan of Daban house in Yong Mausoleum

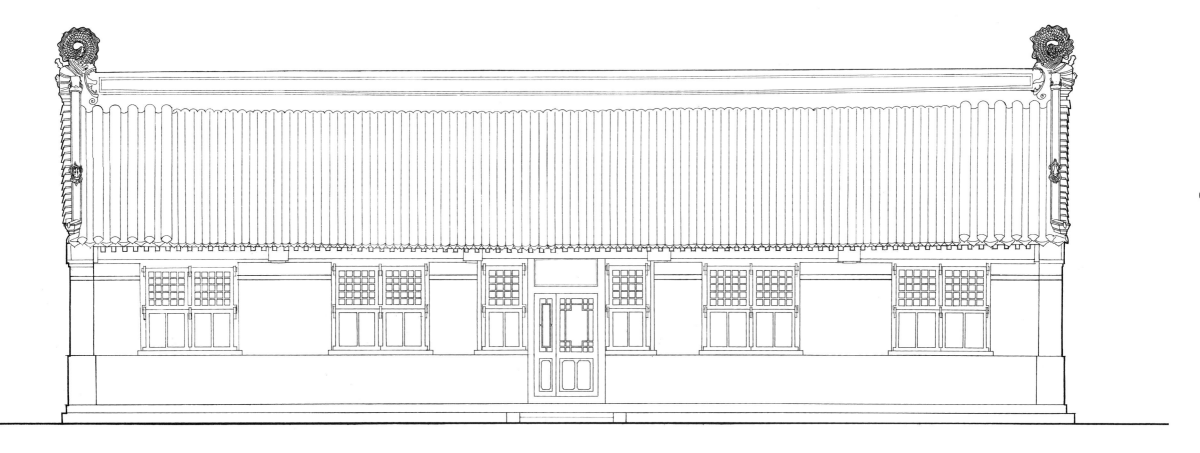

0　0.5　　　3m

清永陵大班房正立面图
Elevation of Daban house in Yong Mausoleum

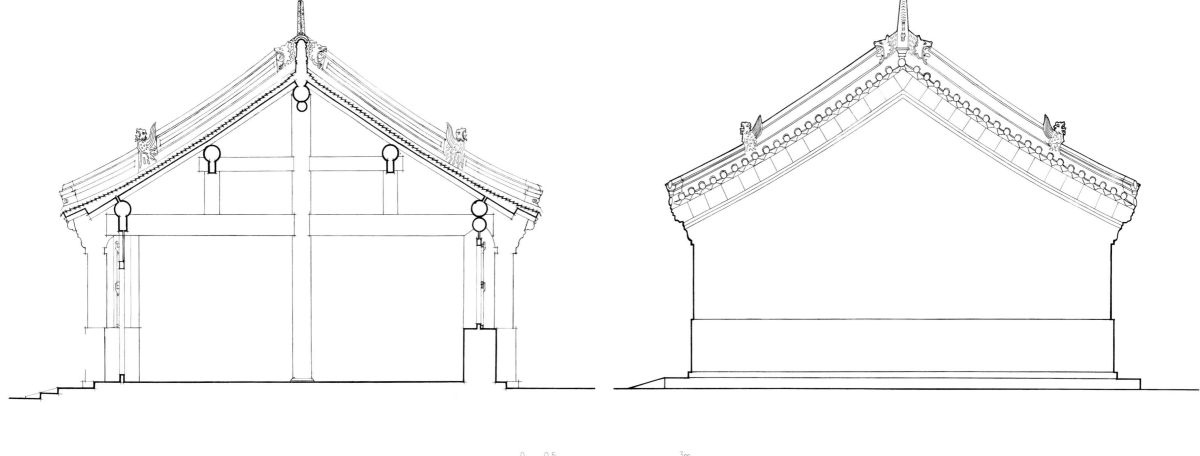

清永陵大班房 1-1 剖面图
1-1 section of Daban house in Yong Mausoleum

清永陵大班房侧立面图
Side elevation of Daban house in Yong Mausoleum

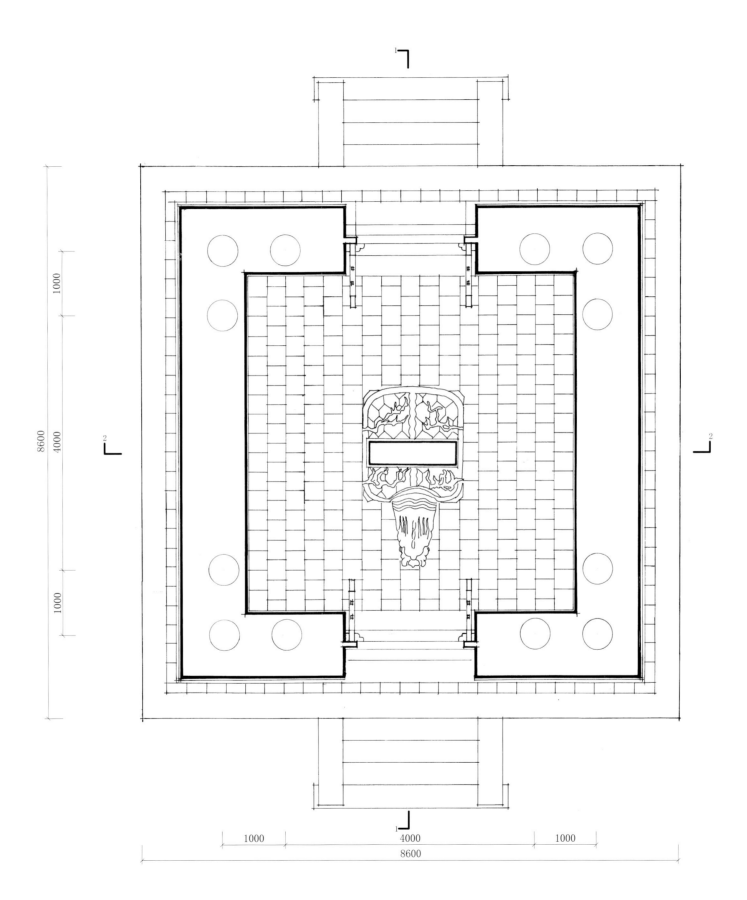

清永陵碑亭平面图
Plan of stele pavilion in Yong Mausoleum

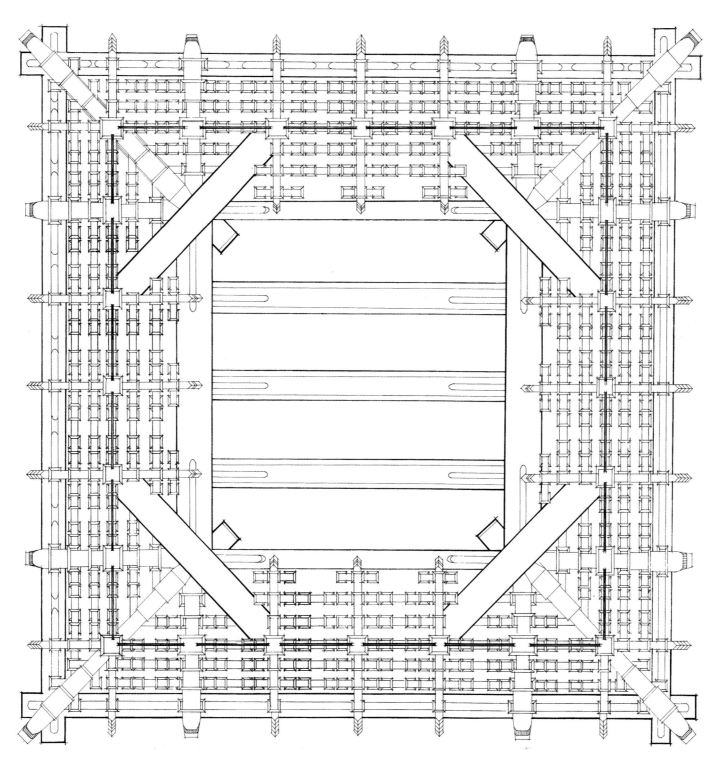

清永陵碑亭梁架仰视图
Beam frame upward plan of stele pavilion in Yong mausoleum

0 15   90cm

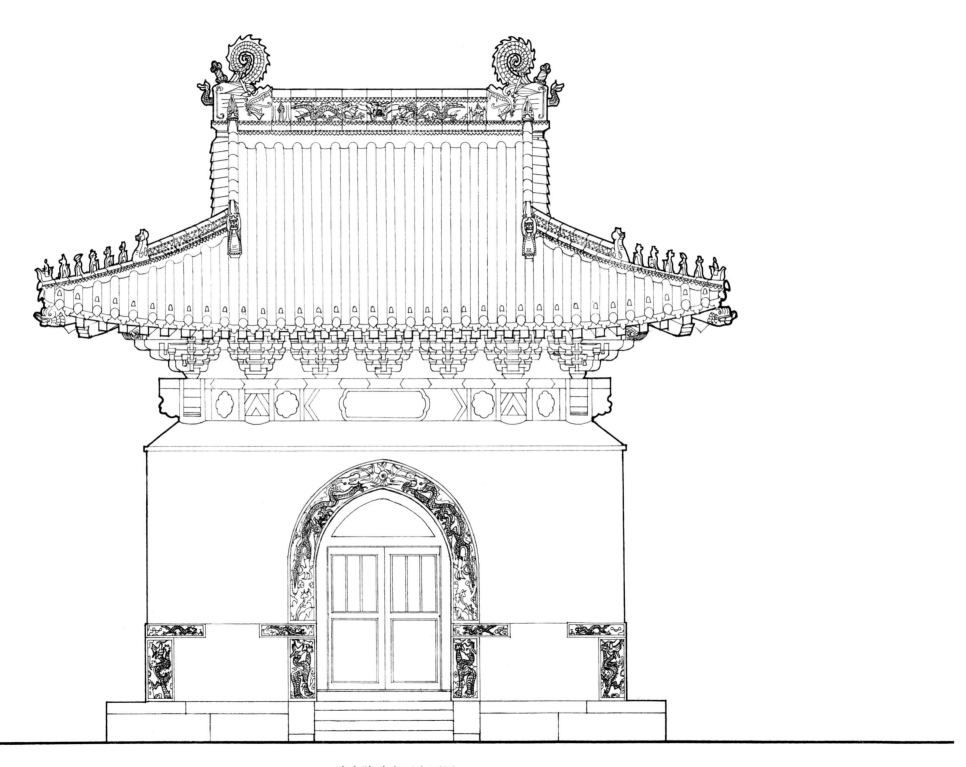

清永陵碑亭正立面图
Elevation of stele pavilion in Yong mausoleum

0   0.5   3m

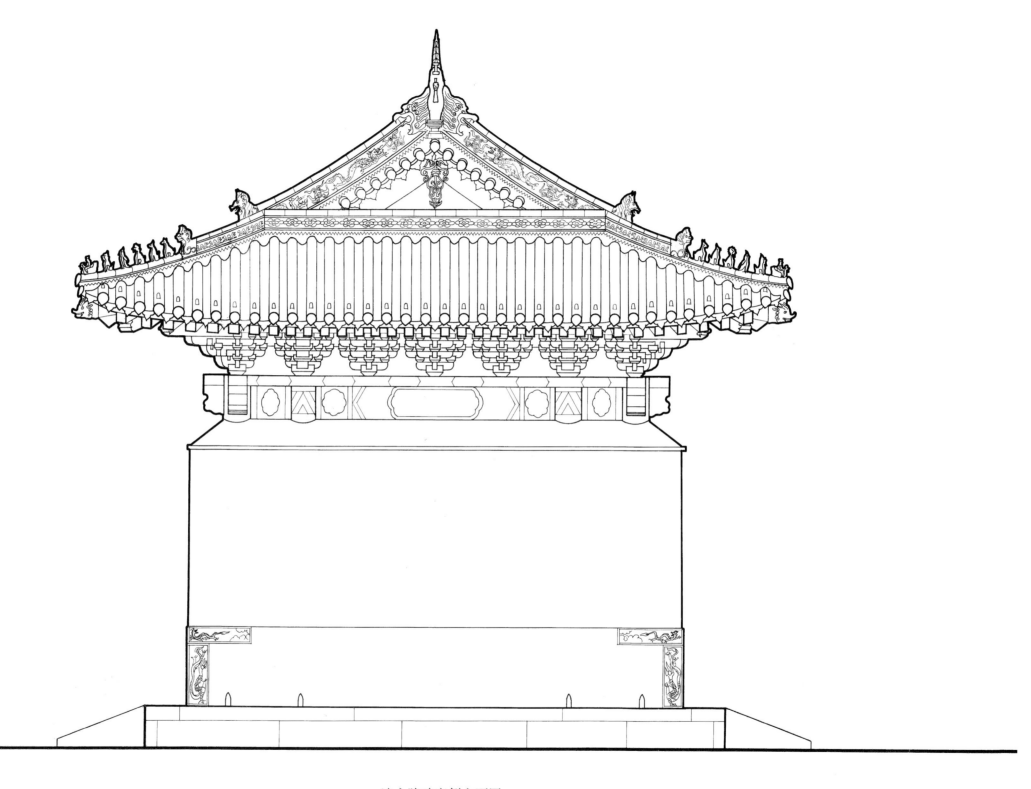

清永陵碑亭侧立面图
Side elevation of stele pavilion in Yong mausoleum

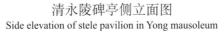

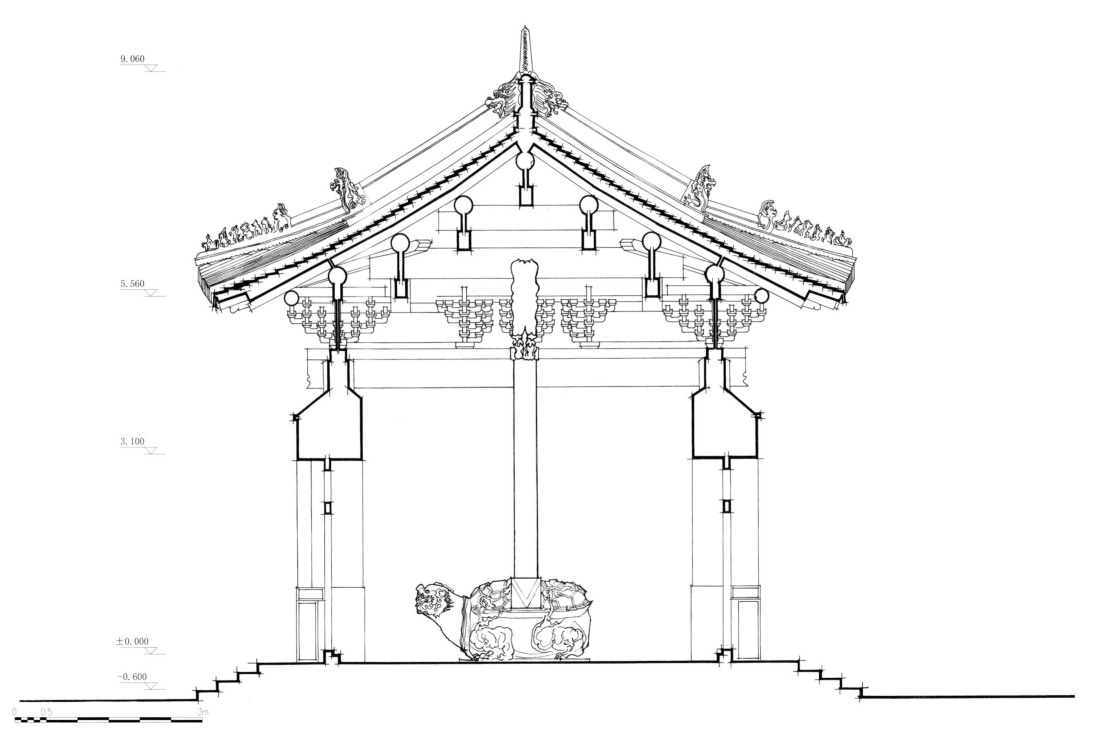

9.060

5.560

3.100

±0.000

−0.600

0   0.5   3m

清永陵碑亭 1-1 剖面图
1-1 section of stele pavilion in Yong mausoleum

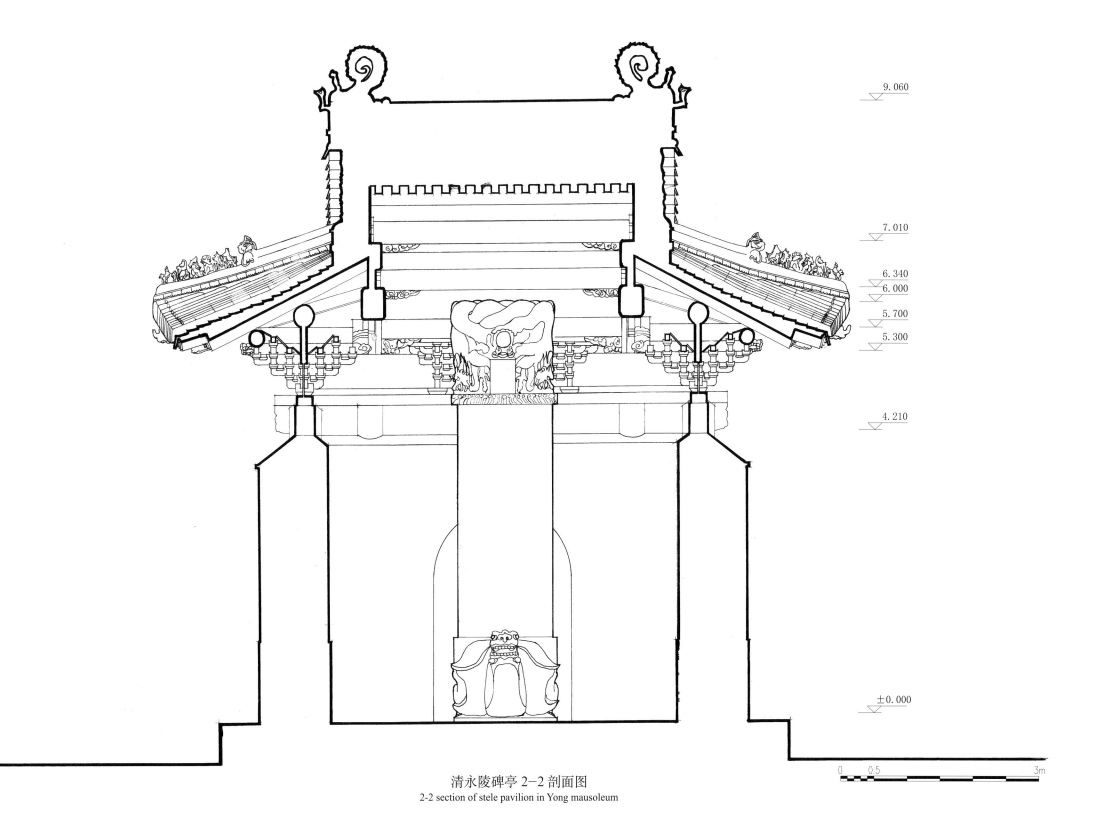

9.060

7.010

6.340
6.000
5.700
5.300

4.210

±0.000

清永陵碑亭 2-2 剖面图
2-2 section of stele pavilion in Yong mausoleum

0   0.5   3m

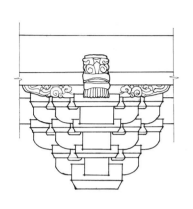

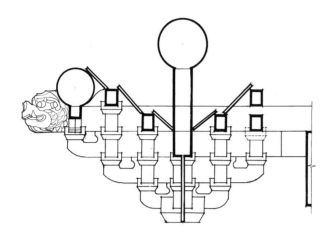

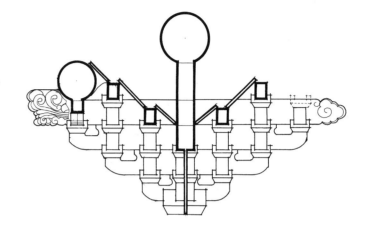

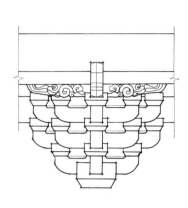

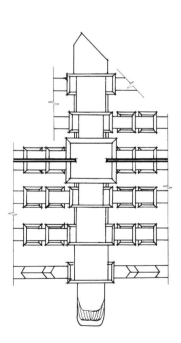

柱头科斗棋主要构件尺寸表（mm）			
构件名称	高	宽	长
坐斗	330	280	155
正心瓜棋	540	120	110
正心万棋	820	125	110
内（外）拽瓜棋	540	80	110
内（外）拽万棋	820	80	110
头翘	570	160	160
二翘	1055	190	160
三翘	1540	220	160
耍头		170	160
槽升子	120	160	80
三才升	120	130	80
十八斗	280	130	80
十八斗	310	130	80
十八斗	310	130	80

平身科斗棋主要构件尺寸表（mm）			
构件名称	高	宽	长
坐斗	250	280	155
正心瓜棋	540	120	110
正心万棋	820	125	110
内（外）拽瓜棋	540	80	110
内（外）拽万棋	820	80	110
头翘	570	80	160
二翘	1055	80	160
三翘	1540	80	160
耍头	2206	80	160
槽升子	120	160	80
三才升	120	130	80
十八斗	150	130	80

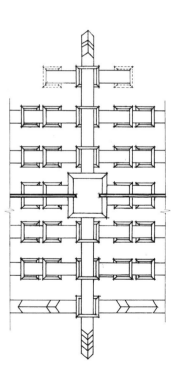

0 10 30cm

清永陵碑亭斗棋大样图（一）

Detail drawings of Dougong on stele pavilion in Yong mausoleum（Ⅰ）

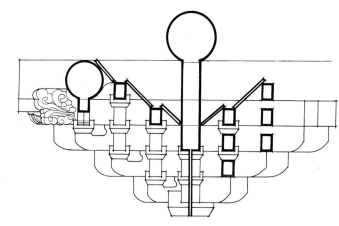

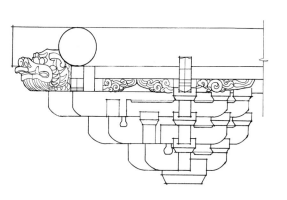

角科斗栱主要构件尺寸表 (mm)			
构件名称	高	宽	长
坐斗	290	290	155
搭角正头翘	275	80	160
斜头翘	880	160	160
搭角正二翘	510	80	160
斜二翘	1570	195	160
搭角正三翘	750	80	160
斜三翘	2250	220	160
外拽瓜栱		80	110
外拽万栱		80	110
十八斗	150	130	80
三才升	120	130	80
槽升子	120	160	80

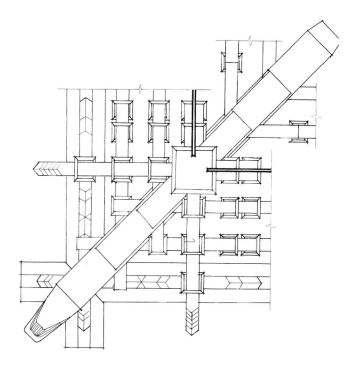

0 10   60cm

清永陵碑亭圣功德碑大样图
Detail drawings of Shenggongde stele in Yong mausoleum

0 10   30cm

清永陵碑亭斗栱大样图（二）
Detail drawings of Dougong on stele pavilion in Yong mausoleum (Ⅱ)

0.5 30m

清永陵碑亭石雕大样图
Detail drawings of stone statues on stele pavilion in Yong mausoleum

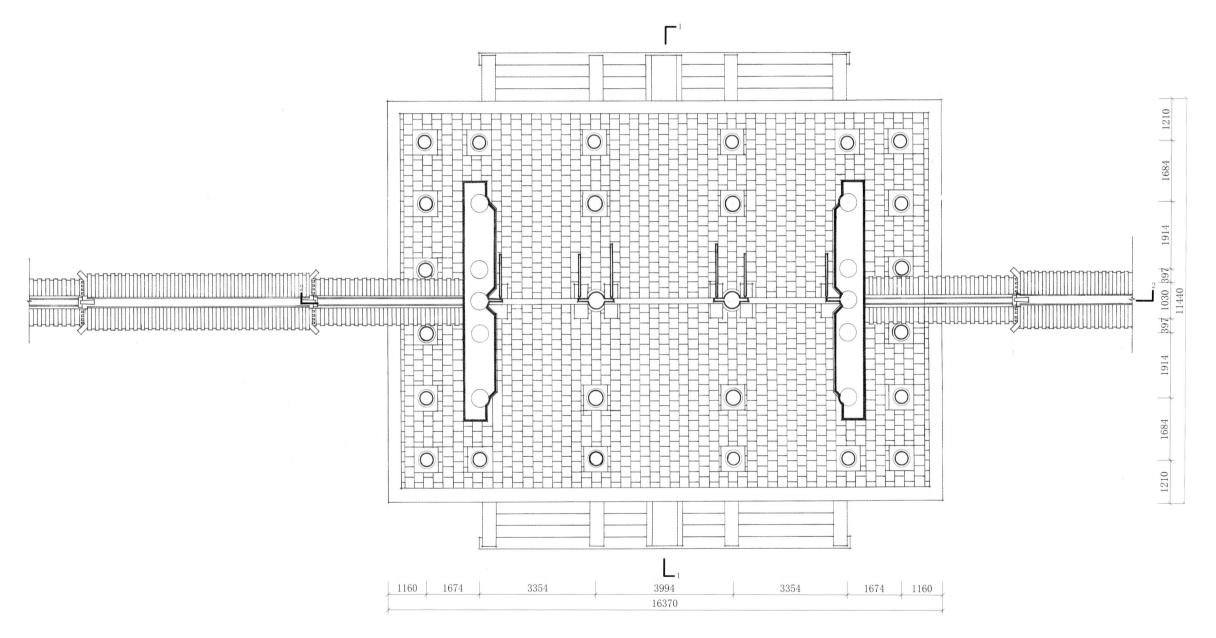

清永陵启运门平面图

Plan of Qiyun gate in Yong Mausoleum

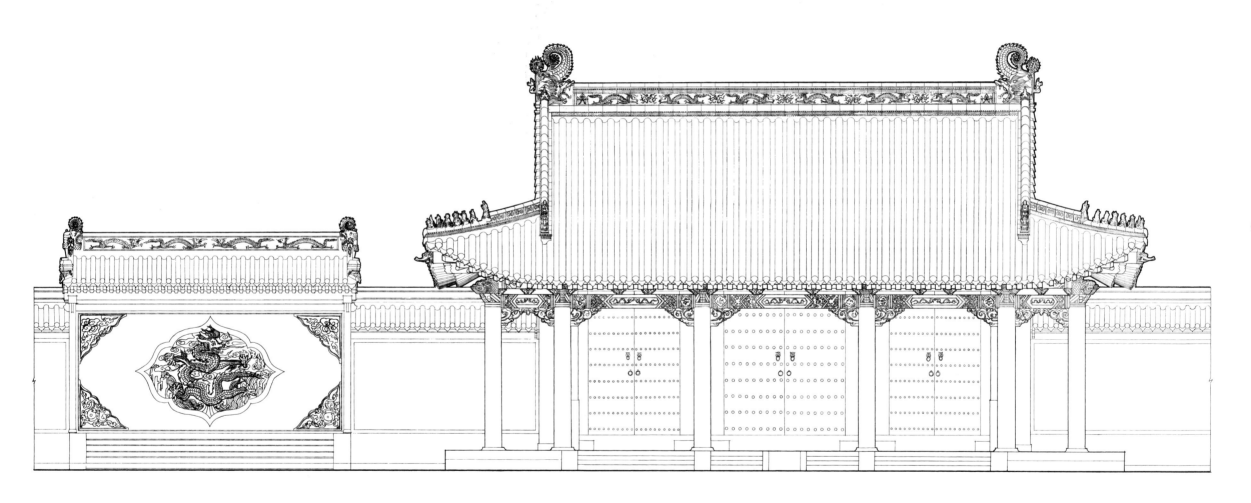

0　0.5　　3m

清永陵启运门正立面图

Elevation of Qiyun gate in Yong Mausoleum

清永陵启运门侧立面图
Side elevation of Qiyun gate in Yong Mausoleum

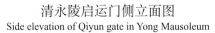

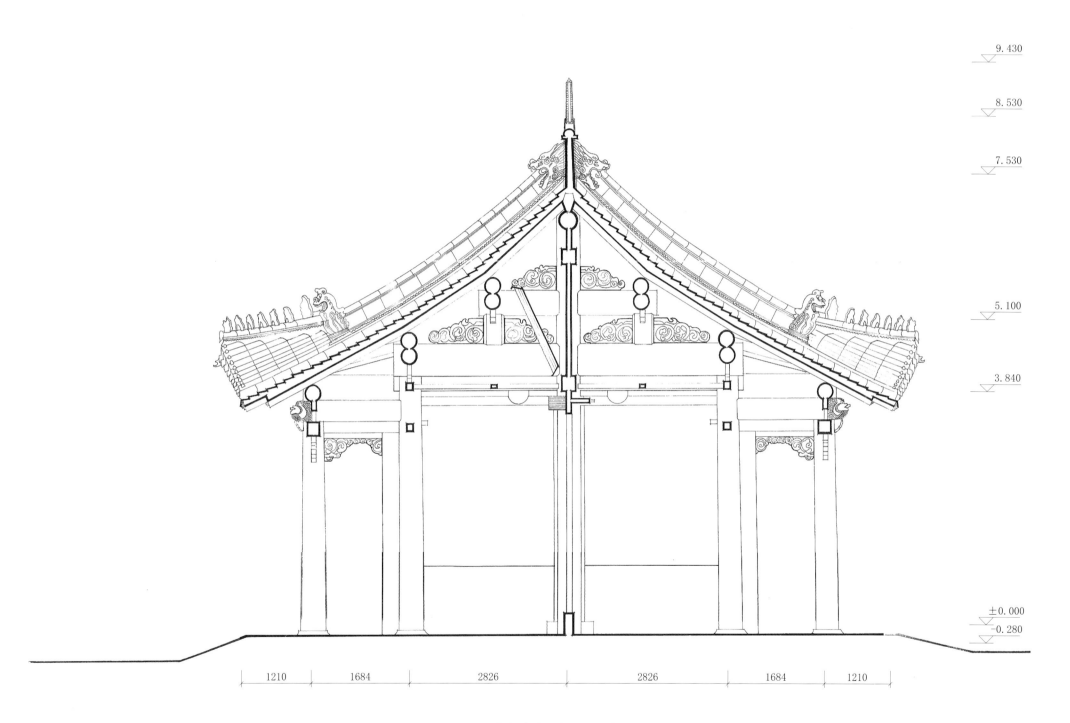

9.430

8.530

7.530

5.100

3.840

±0.000
-0.280

1210　1684　2826　2826　1684　1210

清永陵启运门 1-1 剖面图
1-1 section of Qiyun gate in Yong Mausoleum

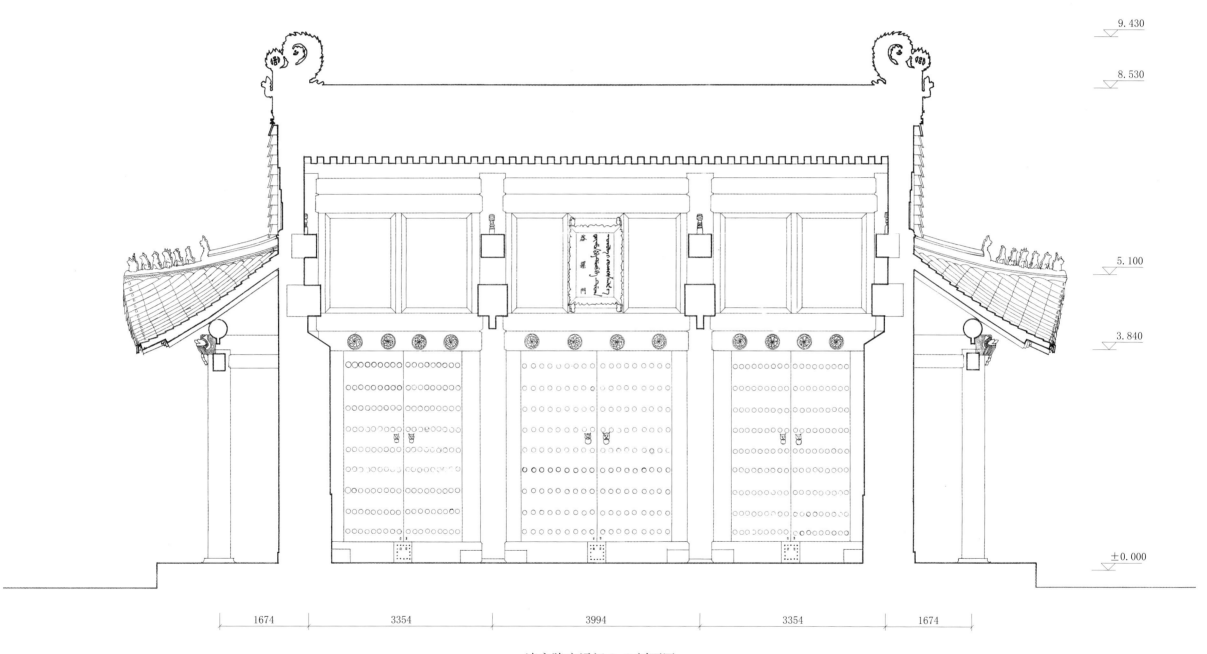

9.430

8.530

5.100

3.840

±0.000

| 1674 | 3354 | 3994 | 3354 | 1674 |

清永陵启运门 2-2 剖面图

2-2 section of Qiyun gate in Yong Mausoleum

清永陵启运门影壁大样图
Detail drawings of screen wall on Qiyun gate in Yong Mausoleum

清永陵启运门牌匾大样图
Detail drawings of name board on Qiyun gate in Yong Mausoleum

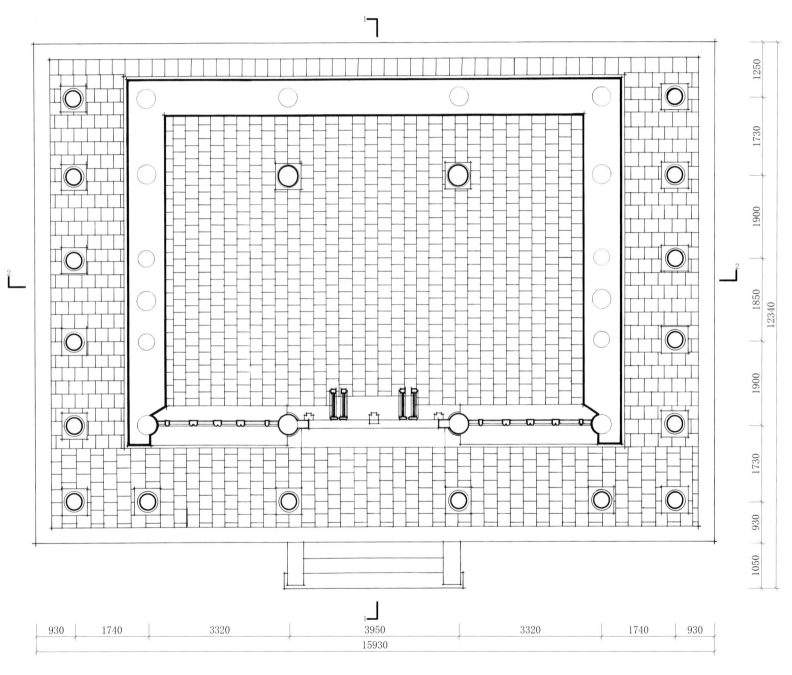

1250
1730
1900
1850
12340
1900
1730
930
1050

930 1740 3320 3950 3320 1740 930
15930

清永陵东配殿平面图
Plan of east supporting hall in Yong Mausoleum

0   0.5   3m

清永陵东配殿正立面图
Elevation of east supporting hall in Yong Mausoleum

清永陵东配殿侧立面图
Side elevation of east supporting hall in Yong Mausoleum

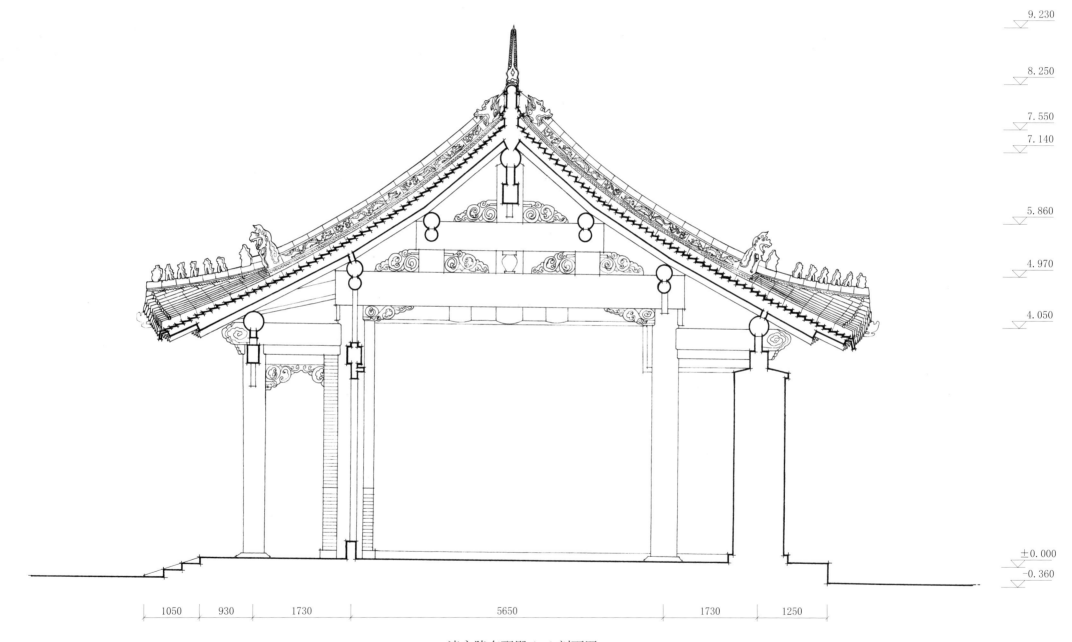

9.230

8.250

7.550

7.140

5.860

4.970

4.050

±0.000

-0.360

| 1050 | 930 | 1730 | 5650 | 1730 | 1250 |

清永陵东配殿 1-1 剖面图

1-1 section of east supporting hall in Yong Mausoleum

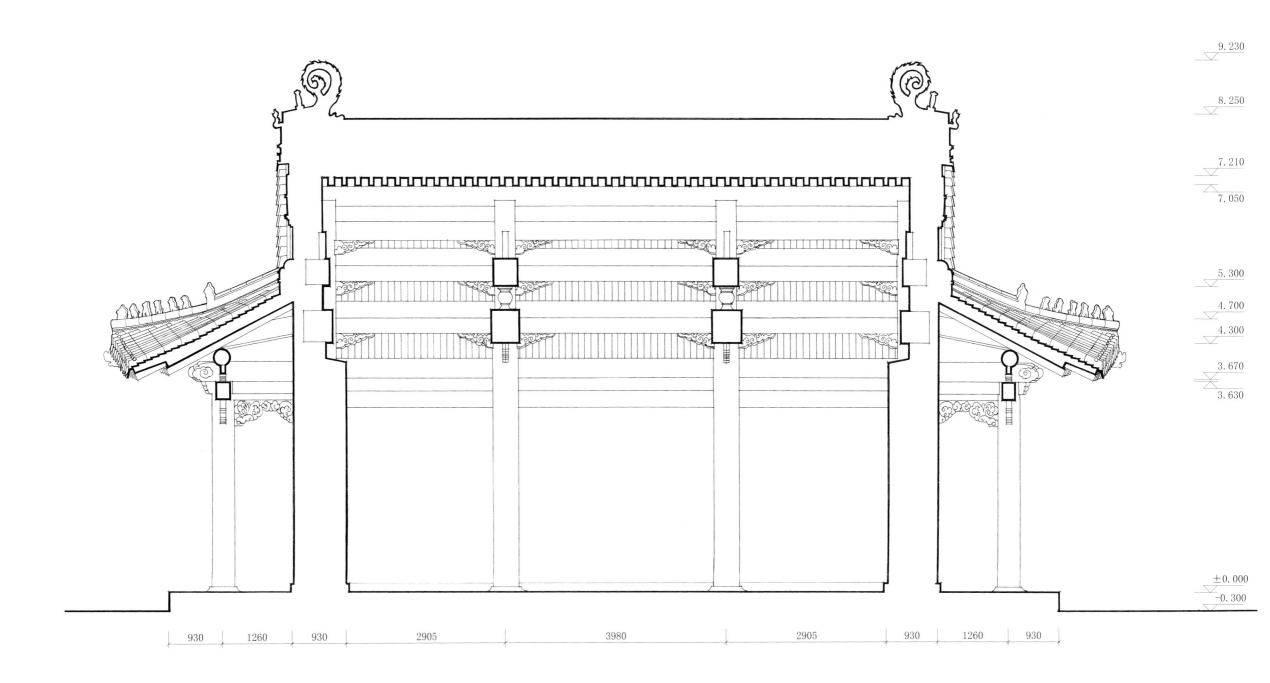

9.230

8.250

7.210

7.050

5.300

4.700

4.300

3.670

3.630

±0.000

−0.300

| 930 | 1260 | 930 | 2905 | 3980 | 2905 | 930 | 1260 | 930 |

清永陵东配殿 2−2 剖面图

2-2 section of east supporting hall in Yong Mausoleum

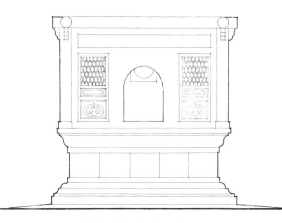

清永陵焚帛炉正立面图
Elevation of stove in Yong Mausoleum

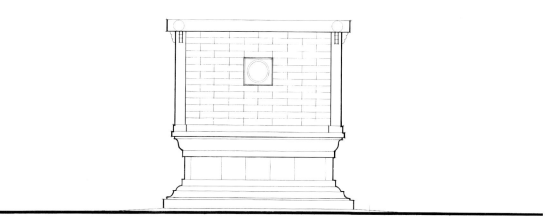

清永陵焚帛炉侧立面图
Side elevation of stove in Yong Mausoleum

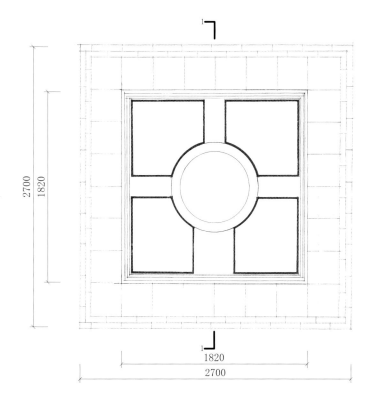

清永陵焚帛炉平面图
Plan of stove in Yong Mausoleum

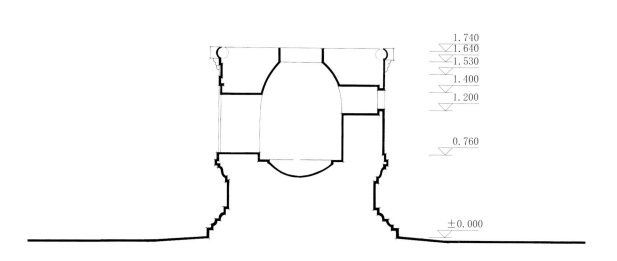

清永陵焚帛炉 1-1 剖面图
1-1 section of stove in Yong Mausoleum

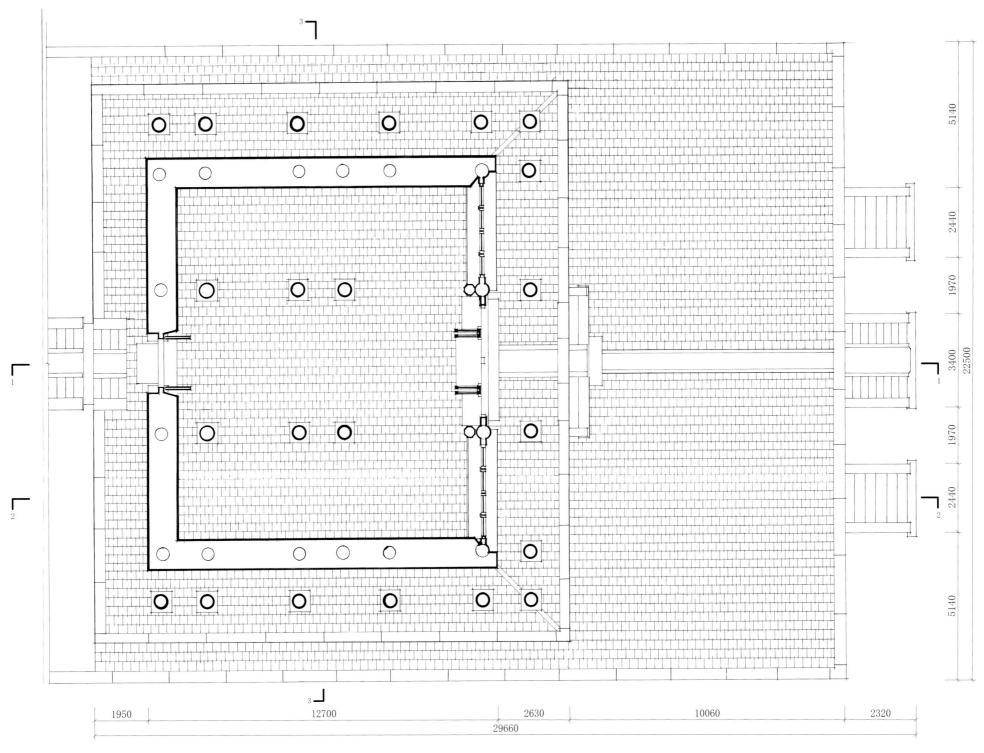

5140

2440

1970

3400

22500

1970

2440

5140

1950
12700
2630
10060
2320

29660

清永陵启运殿平面图
Plan of Qiyun hall in Yong Mausoleum

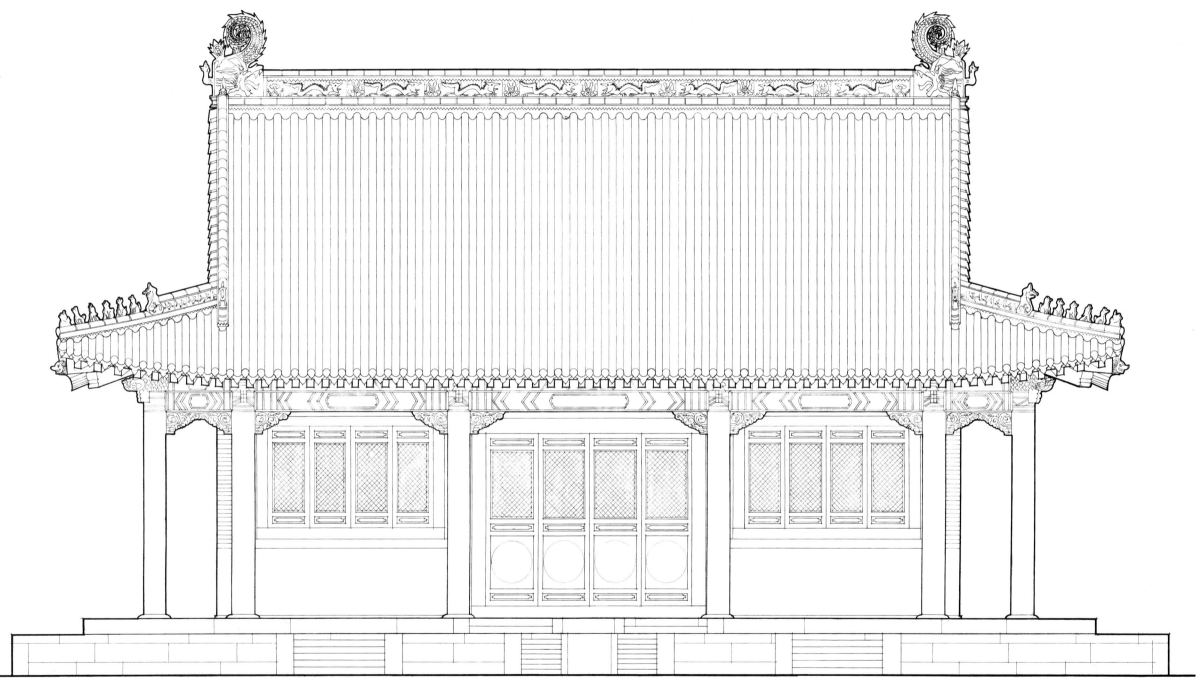

0   0.5   3m

清永陵启运殿正立面图

Elevation of Qiyun hall in Yong Mausoleum

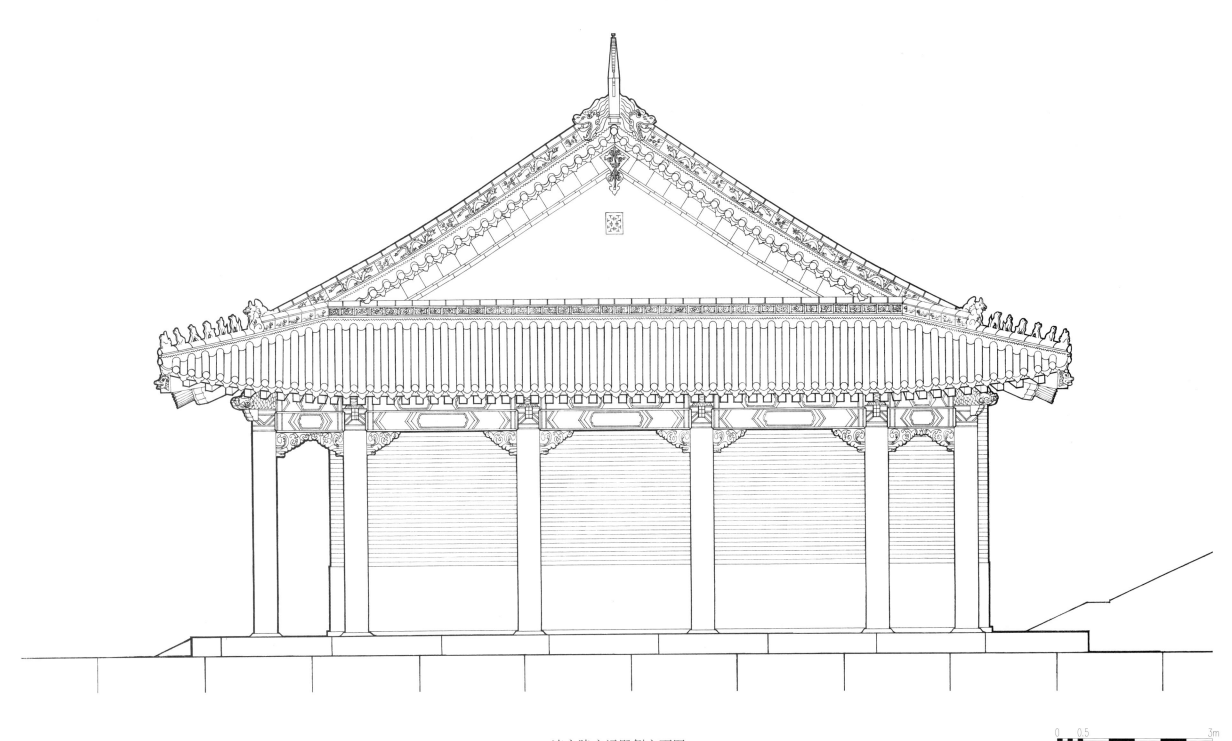

清永陵启运殿侧立面图
Side elevation of Qiyun hall in Yong Mausoleum

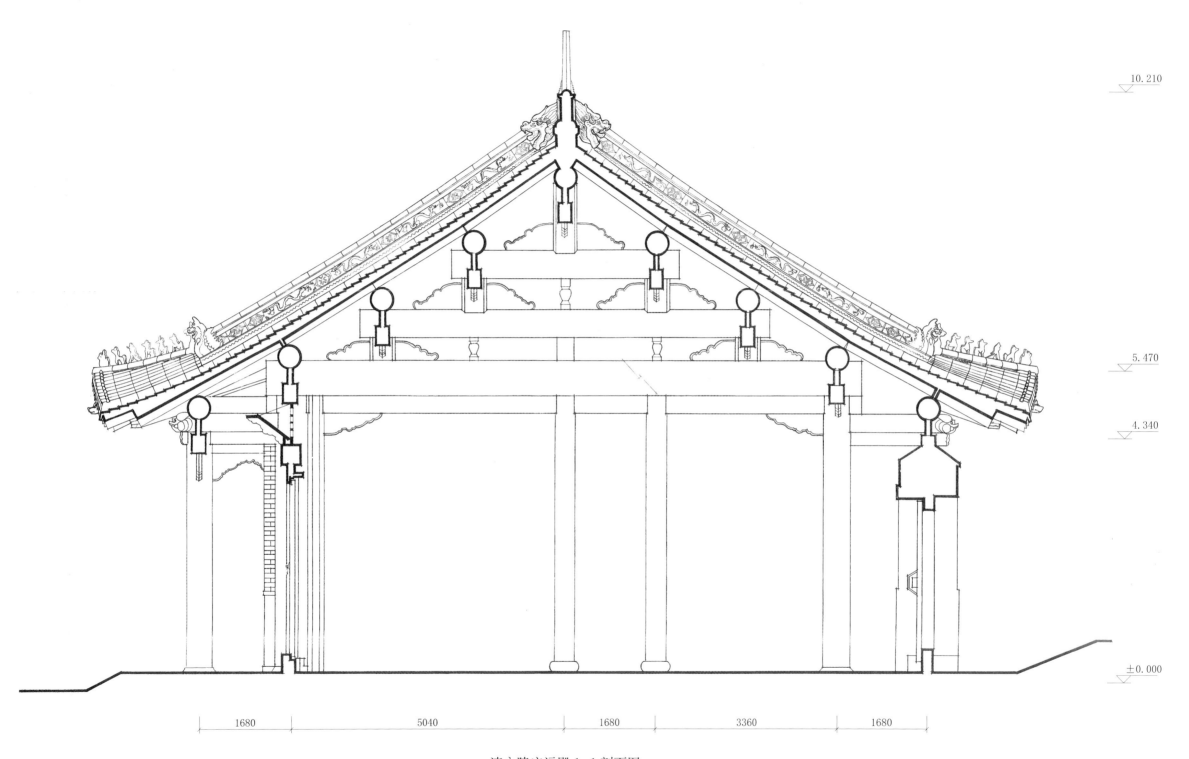

10. 210

5. 470

4. 340

±0. 000

| 1680 | 5040 | 1680 | 3360 | 1680 |

清永陵启运殿 1-1 剖面图

1-1 section of Qiyun hall in Yong Mausoleum

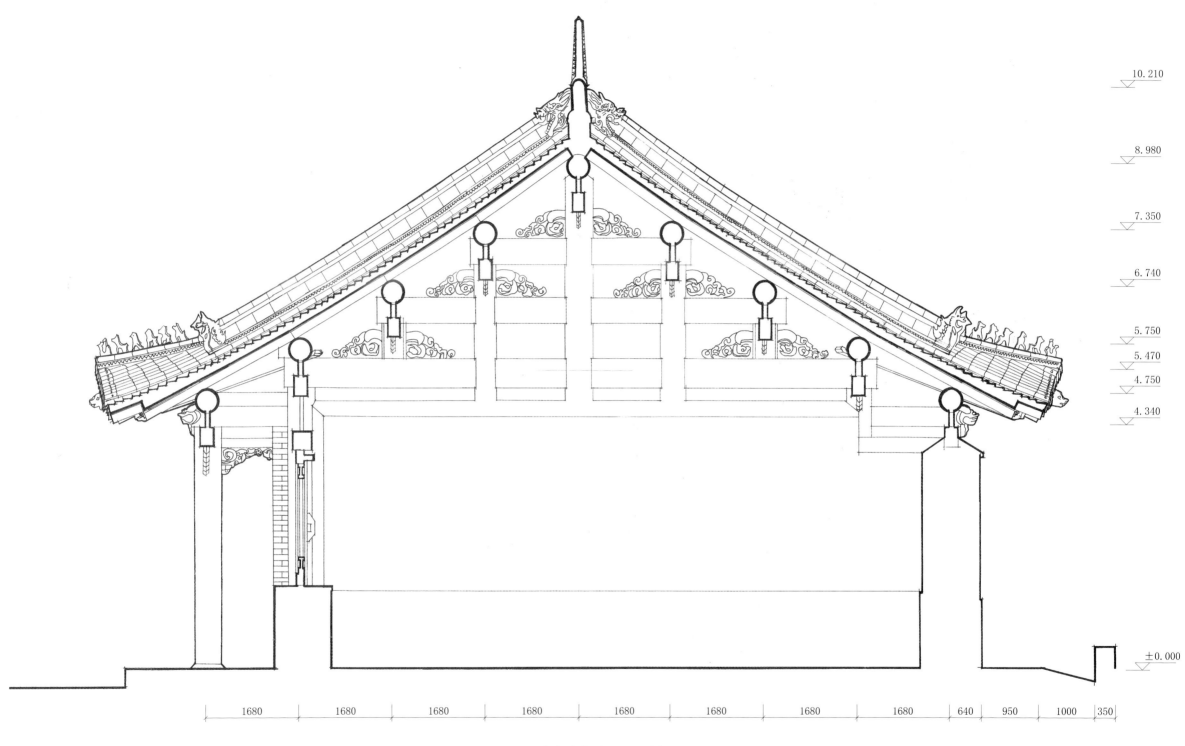

10. 210

8. 980

7. 350

6. 740

5. 750

5. 470

4. 750

4. 340

±0. 000

| 1680 | 1680 | 1680 | 1680 | 1680 | 1680 | 1680 | 1680 | 640 | 950 | 1000 | 350 |

清永陵启运殿 2-2 剖面图

2-2 section of Qiyun hall in Yong Mausoleum

11.310

10.210

8.870

7.900

4.840

4.380

3.000

±0.000

-0.300

-1.080

0   0.5   3m

清永陵启运殿 3-3 剖面图
3-3 section of Qiyun hall in Yong Mausoleum

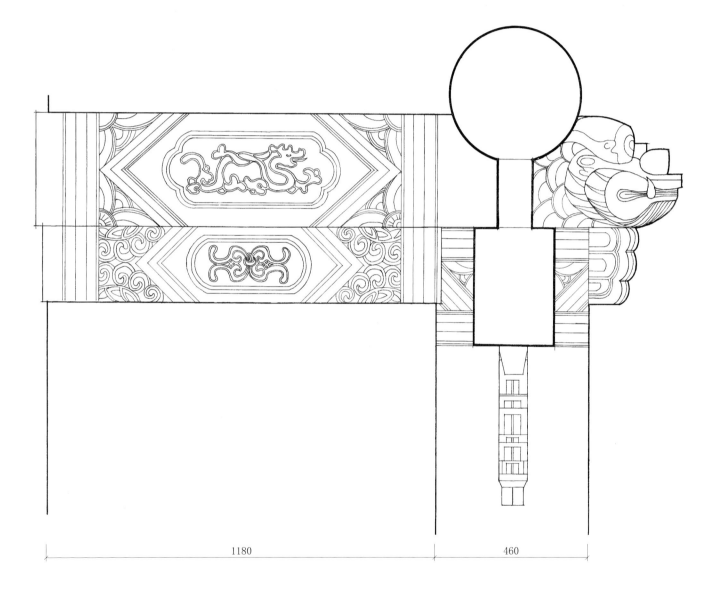

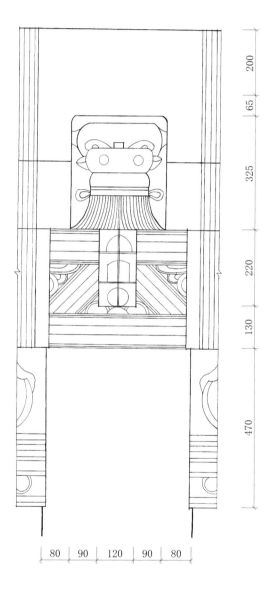

清永陵启运殿抱头梁大样图
Detail drawings of Baotou beam on Qiyun hall in Yong Mausoleum

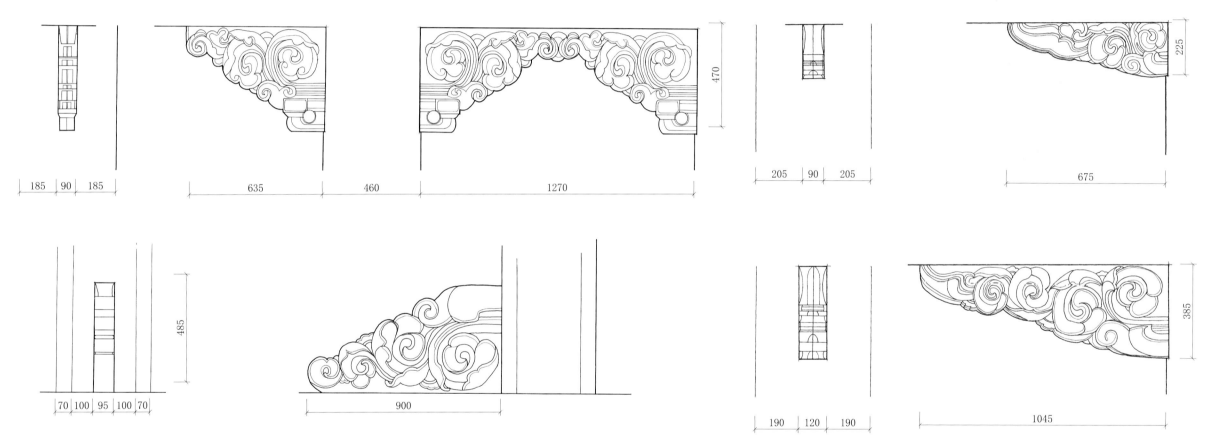

清永陵启运殿雀替角背大样图
Detail drawings of Jiaobei on Queti of Qiyun hall in Yong Mausoleum

清永陵启运殿雀替大样图
Detail drawings of Queti on Qiyun hall in Yong Mausoleum

清福陵

Fu Mausoleum

Project Information

Location: Hunnan district, Shenyang, Liaoning province

Construction Date: 1629, the three year of Tiancong reign period of the Qing dynasty

Area: 194,800 square meters

Administrative Office: Shenyang City Administration and Enforcement Bureau

Responsible Department: Harbin Institute of Architecture and Engineering (Harbin Institute of Technology)

Survey Time: 1985

项目信息

地　　址　辽宁省沈阳市浑南区

始建年代　1629 年（清天聪三年）

占地面积　19.48 万平方米

主管单位　沈阳市城市管理行政执法局

测绘单位　哈尔滨建筑工程学院（现哈尔滨工业大学）

测绘时间　1985 年

1. Brief History

Fu Mausoleum of Qing Dynasty is the Mausoleum for the 1st emperor of Qing Dynasty, Nurhachi and his empress Xiaocigao from Yehenala family, besides, the mausoleum also for his concubine from Wulanalashi family and others. It is known as "Dong Mausoleum" because of its location at the east suburb of Shenyang. It is initially built in the 3rd year of Houjintiancong (the 2nd year of Chongzhen of Ming Dynasty, 1629) and in the 1st year of Chongde named the Mausoleum as "Fu". The mausoleum is finished in the 8th year of Shunzhi (1651), and it is built out and rebuilt in the year of Kangxi, Qianlong and Jiaqing. As the stable of politics and the strengthening of economy during the reign of Emperor Kangxi, the extension of mausoleum put into effect according to institution. The underground palace of the Fu Mausoleum was rebuilt in the 2nd year of Kangxi (1663), where was buried for the funeral urn of the first emperor that set up the throne and memorial tablet in the temple. The tablet of posthumous title was set up in front of the mausoleum in 1664, the Ming tower and the Fangcheng were built in 1665 and 1666, and the monument and pavilion was built in 1688. The two-column archway is built during the reign of Emperor Jiaqing.

2. General Layout

Fu Mausoleum is located at the hillside of Tianzhu Mountain that the topography is raising from south to north, facing Hun River. The continuous mountains are around the Mausoleum with dense trees and the excellent landscape (Fig.1). The cemetery is enclosed by red walls that presents irregular rectangle adapted to the topography, which the length from south to north is 773 meters, the width from east to west is 302 meters, that red gate at both east and west sides are set. There are the boundary marker pillars around the mausoleum area in red, white and blue colors.

The main red gate at the central south is the main entrance of the mausoleum; The section from the north side of Hun River to the main red gate is the guiding part of the mausoleum, along the central axis, where are set a pair of stone stele archways, a pair of ornamental columns and a pair of stone lions on both sides.

When to enter the main red gate, there are four pairs of stone statues in both sides of the central sacred path, including lion, tiger, horse and camel, where a pair ornamental columns is set at front and rear of the stone statues around the dense green pines. The

一、历史沿革

清福陵是清太祖努尔哈赤与孝慈高皇后叶赫那拉氏的陵墓，陵内除了葬有帝后外，还葬有大妃乌喇纳拉氏等后妃。因地处沈阳城东郊，俗称『东陵』。始建于后金天聪三年（明崇祯二年，1629 年），崇德元年（1636 年）定陵号为『福陵』，顺治八年（1651 年）初建完成，康熙、乾隆、嘉庆时期有增建和改建。康熙年间，随着政治稳定，经济实力增强，按制扩建祖陵的工程开始实施。康熙二年（1663 年）改建福陵地宫，葬太祖等人之宝宫于地宫内，在享殿内设宝座神牌。1664 年立谥号碑于陵前，1665 年建明楼，1666 年建方城，1688 年建神功圣德碑及碑亭。嘉庆年间建二柱门。

二、总体布局

福陵布置在自南向北地势渐高的天柱山山坡下，前临浑河，背山面水，陵周峰峦叠嶂，苍翠茂密，景观极佳（图1）。陵园有红墙围合，随地势呈不规则的纵深长方形，南北长约773米，东西宽约302米，东西各设红门一座。陵区四周设有红、白、青三种颜色界桩。

南面正中的正红门为陵园的主入口，从浑河北岸至正红门一段为组群前导，沿中轴线两侧分别布

图2 《盛京风物》载福陵石牌坊及正红门

图1 《盛京风物》载福陵总平面图

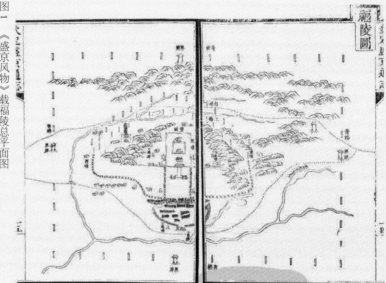

Fig.1 The master plan of Fu mausoleum recorded in *The scenery of Shengjing*

Fig.2 The stone archway and main red gate of Fu mausoleum recorded in *The scenery of Shengjing*

sacred path extends to the north and the mountain steeply raises that a stairs in 108 steps are setup according to the landform, and the stone bridges at both sides of up and down of stairs are built, named Wobo Bridge. The mountain to the north becomes gently. A stele building is built in the centre of sacred path, where the kitchen, tearoom at east and the washroom and fruit room at west are built between the stele pavilion and Fangcheng.

The main part of cemetery is Fangcheng shaped like a castle, that set turrets around and gate tower at the south and north. The centre of south gate wall is Longen Gate and the north is Ming Tower, where the Longen Hall is located at the inner axis of Fangcheng and closes to the Ming Tower, the side halls are set at both sides in front of the hall. The rear of Longen Hall is two-column archway and five stone wares for offerings, and then is vaulted door. Entering the vaulted door under the Ming tower is the Crescent Courtyard which frontage is glazed screen wall and there is a step path on both sides, who can walk up and down to the Fangcheng. Baocheng is behind the Crescent Courtyard where has the bulged Baoding. The underground palace is inside the Baoding.

3.Individual buildings

1) The stone archway

The stone archway in the type of Chongtian has four columns, three bays and three roofs, which erected both east and west sides in front of the main red gate, that is caved on deep grey stone on whole. Columns are all with square and splayed sections, dropping on the pinch stone that is held by drum-shaped bearing stones at front and rear, where has overlapping lotus seat and beasts on the top of column. The surface of the pinch stone, architrave and floral board are full of high relief. The small architrave is higher than the big architrave; each big architrave bears one whole and two half Wucaipinzi Dougong between columns. The three roofs are all overhanging gable roof with large eaves (Fig.2).

2) The main red gate

The main red gate is a three-hole arched door with yellow glazed roof tile, gable and hip roof with single eave. It locates on the Xumi stylobate without Guijiao layer, of which three stairs is set on both front and rear. The voussoir, the stone of slab and corner pier are

三、单体建筑

1. 石牌楼

石牌楼为四柱三间三楼柱冲天式，分立在正红门前方的东西两侧，通体深青色石材雕成。柱子均为方形抹角断面，下落于夹杆石上，夹杆石前后有抱鼓石夹持，柱顶有仰覆莲座，上有望兽。夹杆石与额枋、花板表面布满高浮雕。小额枋的高度大于大额枋，每个大额枋上承托一个整攒、两个半攒平身科五踩品字斗栱。三楼均为悬山式屋顶，出檐较大（图2）。

置石牌楼一对，华表一对，石狮一对。

进入正红门，中央神道的两侧排列着狮、虎、马、驼四对石象生，石象生前后各耸立一对华表，周围是茂密的苍松。神道向北延伸，山势陡升，结合地形设置了俗称『一百〇八蹬』的108级台阶，台阶上下两端修建石桥，称卧波桥。再向北山势变缓，神道正中建碑楼一座，碑楼与方城之间东侧设膳房、茶房，西侧设涤器房、果房。

陵寝的主体是城堡式的方城，四周设角楼，南北设城楼，城南墙正中是隆恩门，北墙正中是明楼，隆恩殿位于方城内轴线北部靠近明楼处，殿前两侧有配殿。隆恩殿后有二柱门和石五供，再后是券门。

步入明楼下的券门就进入了月牙城，月牙城正面有琉璃影壁，两侧有『蹬道』可上下方城，月牙城之后是宝城，内为隆起的封土宝顶，宝顶之内为地宫。

图4 《盛京风物》载福陵方城明楼

图3 《盛京风物》载福陵碑亭

Fig.3 The stele powilion of Fu mausoleum recorded in *The scenery of Shengjing*
Fig.4 The Longen hall and five stone tributes of Fu mausoleum recorded in *The scenery of Shengjing*

all full of dragon in high relief. There are many components such glazed flying rafter, eave rafter, Dougong, architrave, engraving partition, dropping floral columns and so on under the eave. The intervals of dropping floral columns is the same and each interval is used three sets of Wucaipinzi Dougong between the columns. There are the sleeve walls on both sides of the gate, the center of which a colored glaze twisted dragon is filled in (See Fig.2).

3) The stele pavilion

The stele pavilion with square plane and side length in 9.76 meters has gable and hip roof of double eaves and is covered by yellow glazed tile. The body of wall is made of brick with a vaulted door on each side, that is obviously narrowed on upper part and the bottom of wall is up right. The lower eave uses Chong'ang Wucai Dougong, the upper eave uses Danqiao Chong'ang Qicai Dougong. Exposed components like the architrave are decorated in Xuanzi colored painting. The Xumi stylobate is without decorations and four sides are built a stone stair with side dropping belt stone, which a platform is added under the stylobate. The "Daqingshengong" sacred stele is located in side the pavilion in height of 6.67 meters, width of 1.8 meters and depth of 0.72 meters (Fig.3).

4) Fangcheng

Fangcheng is built with black bricks, which circumference is 379.5 meters, wall height is 5.23 meters and crenel height is 1.67 meters. The inner side of Fangcheng is parapet and the berm on wall is 2 meters wide (Fig.4).

5) The Longen Gate

The Longen Gate is located on a pier with 15.5 meters wide, 12.1 meters deep and 5.65 meters high. The central vaulted door is out of back of the wall and only the upper part of the voussoir exposed. The upper part of the voussoir is built like a sunk panel, which the center is gate plaque with floral decorations on four corner. The pier is without belt line stone and the upper and tip of pier are as a whole. The gatehouse has three storeys with three bays wide and two bays deep, surrounding by peripteral each storey. The width of the central bay is the same at each storey, but the width of side rooms reduces from bottom to top in proper. The gable and hip roof and two-storey eaves are covered by yellow glazed tile, the diagonal ridge for gable and hip roof of top storey and the angle ridges of eaves are all decorated by three running beasts but no immortal (Fig.5).

2. 正红门

正红门为单檐歇山黄琉璃瓦顶三孔券门。坐落在无圭角层的须弥座台基上，前后各出三路垂带踏跺。各间券脸石与压面石、角柱石均布满高浮雕龙纹。檐下有琉璃的飞椽、檐椽、斗栱、额枋、花罩、垂花柱等构件，垂花柱间距相等，每开间均用三攒平身科五踩品字斗栱。门两侧做袖壁，中心嵌五彩琉璃蟠龙（见图2）。

3. 碑亭

碑亭平面为正方形，每边长9.76米。重檐歇山顶，上覆黄色琉璃瓦。墙体上身有明显收分，下碱直立。下檐用重昂五踩斗栱，上檐用单翘重昂七踩斗栱。额枋等露明构件饰旋子彩画。台基为素面须弥座，四面各出一路垂带踏跺，台基下加做一层月台。亭内『大清神功圣德碑』通高6.67米，碑身宽1.8米，厚0.72米（图3）。

4. 方城

方城为青砖砌筑，周长379.5米，城墙高5.23米，垛口高1.67米，内侧做女儿墙，墙上马道宽2米（图4）。

5. 隆恩门

隆恩门坐落在宽15.5米、深12.1米、高5.65米的墩台上。正中的券洞门退到城墙外皮之后，券脸石只露出上半部分。券脸石的上方做成方池子，四个岔角有花饰，中心为门匾。墩台无腰线石，上身与下碱联成一体。门楼高三层，各层均面阔三间、进深两间、带周围廊。各层明间面阔相同，次间面阔则自下而上依次缩减。歇山屋顶与两层腰檐均用黄色琉璃瓦，顶层戗脊与各腰檐角脊均置三个跑兽，无仙人（图5）。

图6 《盛京风物》载福陵隆恩殿及石五供

图5 《盛京风物》载福陵隆恩门

Fig.5　The Longen gate of Fu mausoleum recorded in *The scenery of Shengjing*
Fig.6　The Fangcheng and Minglou of Fu mausoleum recorded in *The scenery of Shengjing*

6) The Longen Hall

The Longen Hall is also named "Xiang Hall", "Shen Hall", where is the main mausoleum and important worship were held. It uses gable and hip roof with single eave in three bays wide and two bays deep, surrounding by peripteral. The pedestal of the hall is low and is the simplify Xumi stylobate, where the large platform is under the pedestal around stone railings; the southern of the platform has three stairs and the front stair is the imperial step with the embossment of two dragons playing with a pearl. The central bay and side bays uses two sets of San'ang Qicai Dougong between the columns, where the upper of eave column is attached three beasts' head decorations. The beam frames are used in corridor type on both front and back of nine purlins, which central columns are added under the six step beams of the two trusses of the central bay. The interior of hall is without ceiling that the interior and exterior of the hall are painted Xuanzi paintings. The angle ridges are decorated with five beasts and without immortal. A shrine is set in the back of the central bay as a warmroom, which contains altar to sacrifice the owner of mausoleum (Fig.6).

7) The two-column archway

The two-column archway is a stele archway with single bay and one roof with two towering columns. Two towering columns are square and the tip of columns is clamped by a pedestal with rolling stone at front and back, the top of columns is square Xumi stylobate with the squatted beast facing north. The tie beam is wooden construction and painted Xuanzi painting, and ten sets of Qicai Pinzi Dougong between the columns are built on the flat tie beam. The top is overhanging gable roof and there is a larger space between the gable eave board and towering column. There is a door with two partitions at center and both sides have large boards between the door leaf and column.

参考文献
References

[1] 王伯扬．中国古建筑大系（第 2 册）：帝王陵寝建筑 [M]．北京：中国建筑工业出版社，1993．

[2] 潘谷西．中国古代建筑史（第四卷）：元明建筑（第二版）[M]．北京：中国建筑工业出版社，2009．

[3] 陈伯超，刘大平，李之吉．中国古代建筑史全集（辽黑吉卷）[M]．北京：中国建筑工业出版社，2016．

[4] 村田治郎．亚细亚大观 [J]．亚细亚写真大观，1942．

[5] 侯幼彬．读建筑 [M]．北京：中国建筑工业出版社，2012．

[6] 辽宁省图书馆．盛京风物 [M]．北京：中国人民大学出版社，2006．

[7] 黑龙江省博物馆．中东铁路大画册 [M]．哈尔滨：黑龙江人民出版社，2013．

6. 隆恩殿

又称『享殿』『神殿』，是供奉陵主神位和举行祭祀的重要场所。单檐歇山顶，面阔三间，进深两间，带周围廊。殿身台基低矮，为简化的须弥座，台基下面的大月台尺度庞大，环砌石栏杆，南向三出陛，正阶为御路踏跺，御路石雕二龙戏珠。明间、次间均用两攒斗口三昂七踩平身科斗栱，檐柱上端均附三幅云兽头花饰。九檩前后廊式梁架，明间两缝七架梁下加立中柱。殿身内彻上明造，殿内外皆用旋子彩画。戗脊上走兽数为五个，无仙人。明间后部设佛龛式暖阁，阁内置神座，供奉陵主（图 6）。

7. 二柱门

二柱门为单间二柱一楼冲天式牌楼门。两根冲天柱为方形，下端被前后带滚礅石的基座夹持，顶部有方形须弥座，上置背南面北蹲坐的望兽。梁枋均为木构，用旋子彩画，平板枋上施十攒七踩品字斗栱。楼顶为悬山式，博缝板与冲天柱之间有较大空挡。中间开启两扇槅扇门，两侧是宽大的余塞板。

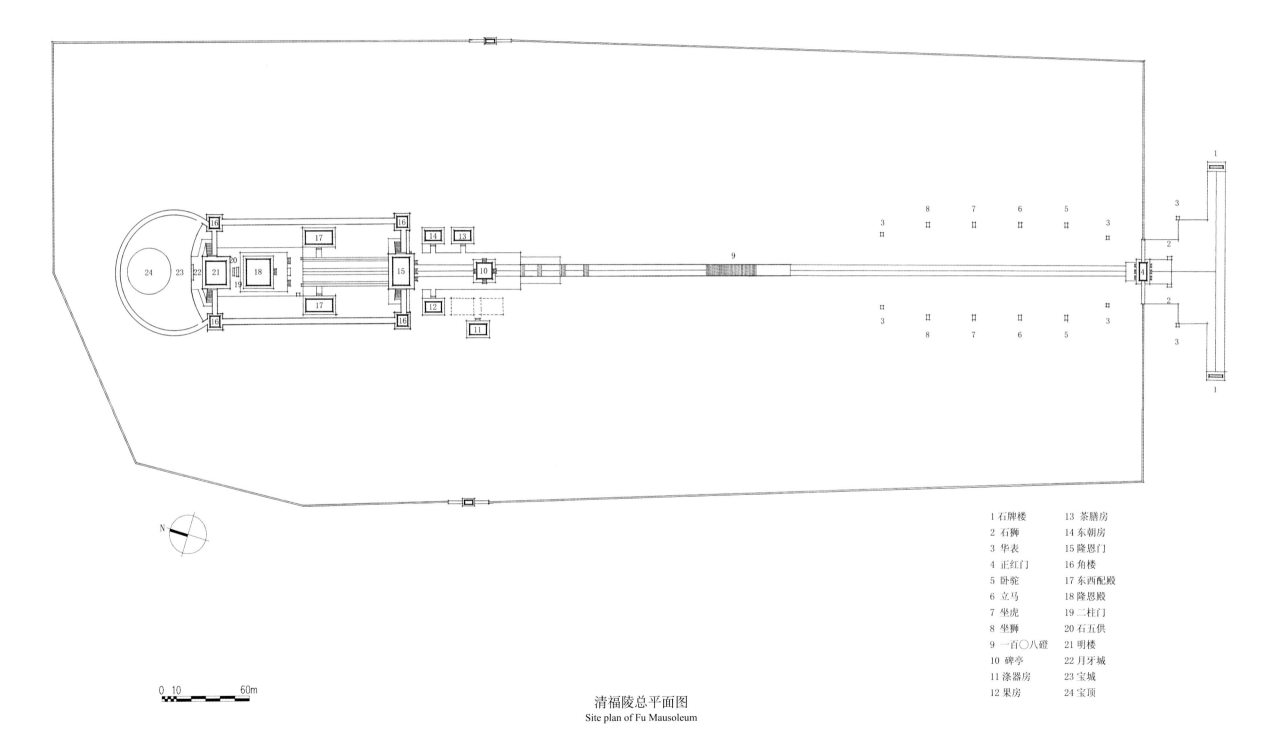

N

0 10 60m

清福陵总平面图
Site plan of Fu Mausoleum

1 石牌楼	13 茶膳房
2 石狮	14 东朝房
3 华表	15 隆恩门
4 正红门	16 角楼
5 卧驼	17 东西配殿
6 立马	18 隆恩殿
7 坐虎	19 二柱门
8 坐狮	20 石五供
9 一百〇八磴	21 明楼
10 碑亭	22 月牙城
11 涤器房	23 宝城
12 果房	24 宝顶

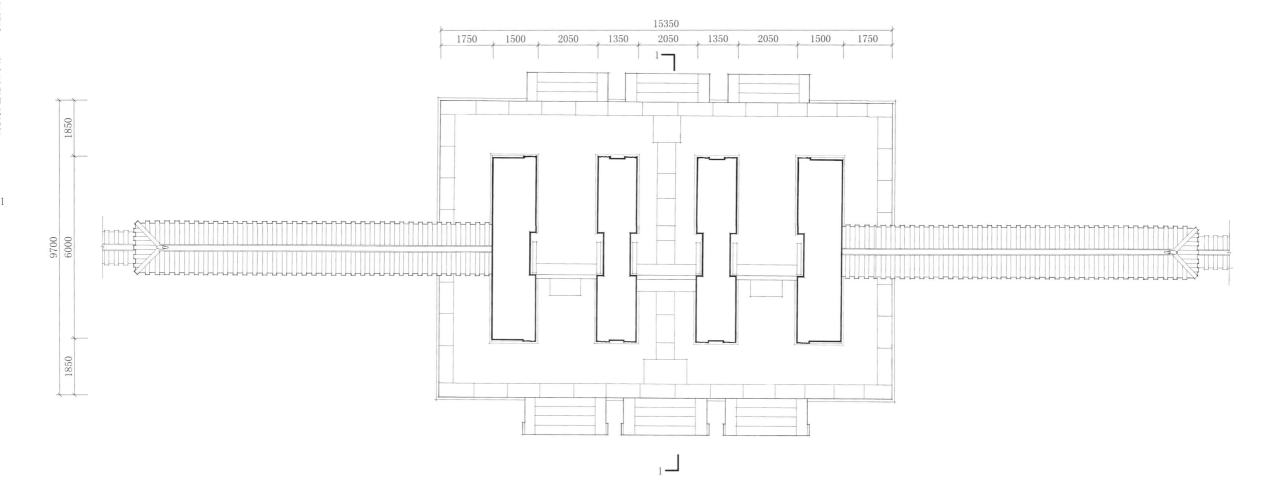

清福陵正红门平面图
Plan of main red gate in Fu mausoleum

清福陵正红门正立面图
Elevation of main red gate in Fu mausoleum

0   0.5   2m

清福陵正红门侧立面图
Side elevation of main red gate in Fu mausoleum

0   0.5   2m

8.300

5.600

±0.000

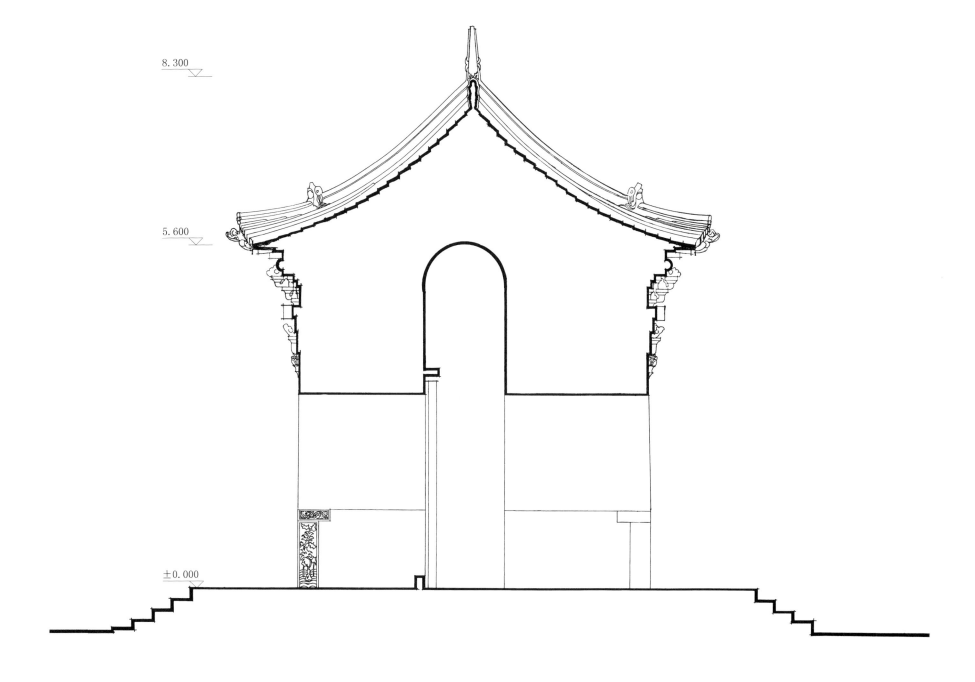

清福陵正红门 1—1 剖面图

1-1 section of main red gate in Fu mausoleum

0   0.5   2m

清福陵正红门袖壁大样图
Detail drawings of sleeve wall beside main red gate in Beizhen temple

0   0.5   1m

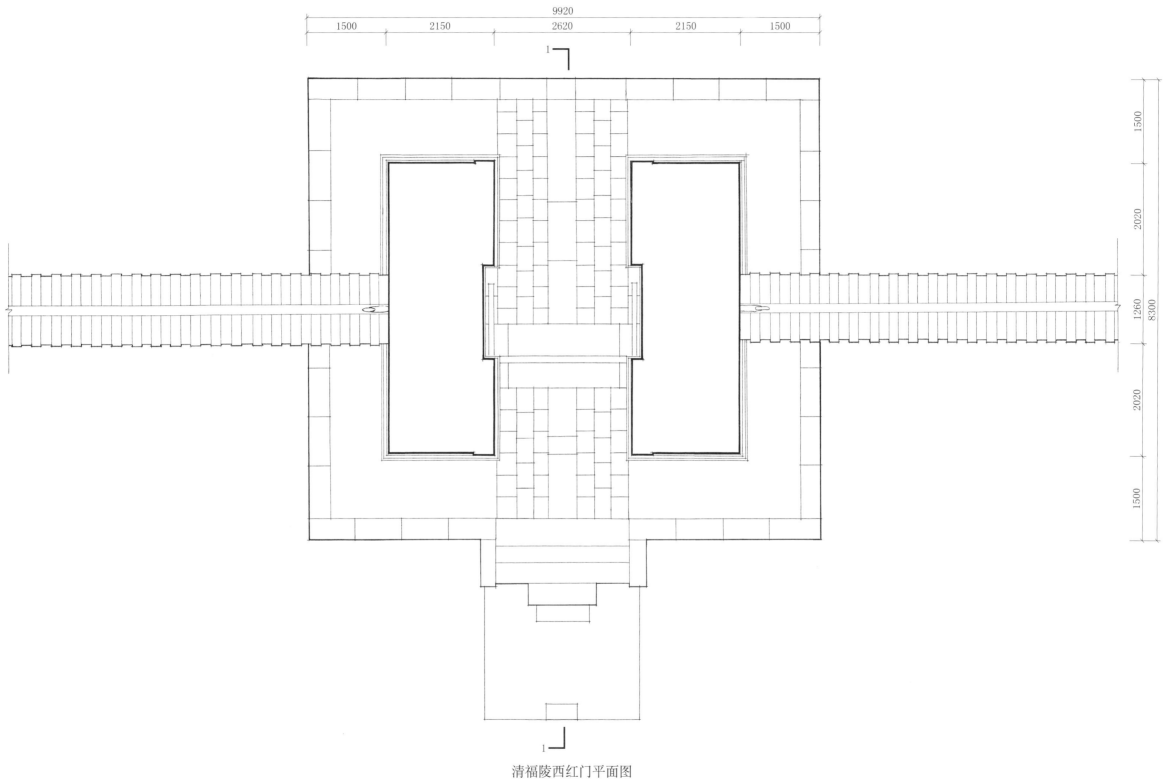

清福陵西红门平面图
Plan of west red gate in Fu mausoleum

0   0.5   2m

清福陵西红门正立面图
Elevation of west red gate in Fu mausoleum

清福陵西红门侧立面图
Side elevation of west red gate in Fu mausoleum

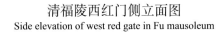

0   0.5   2m

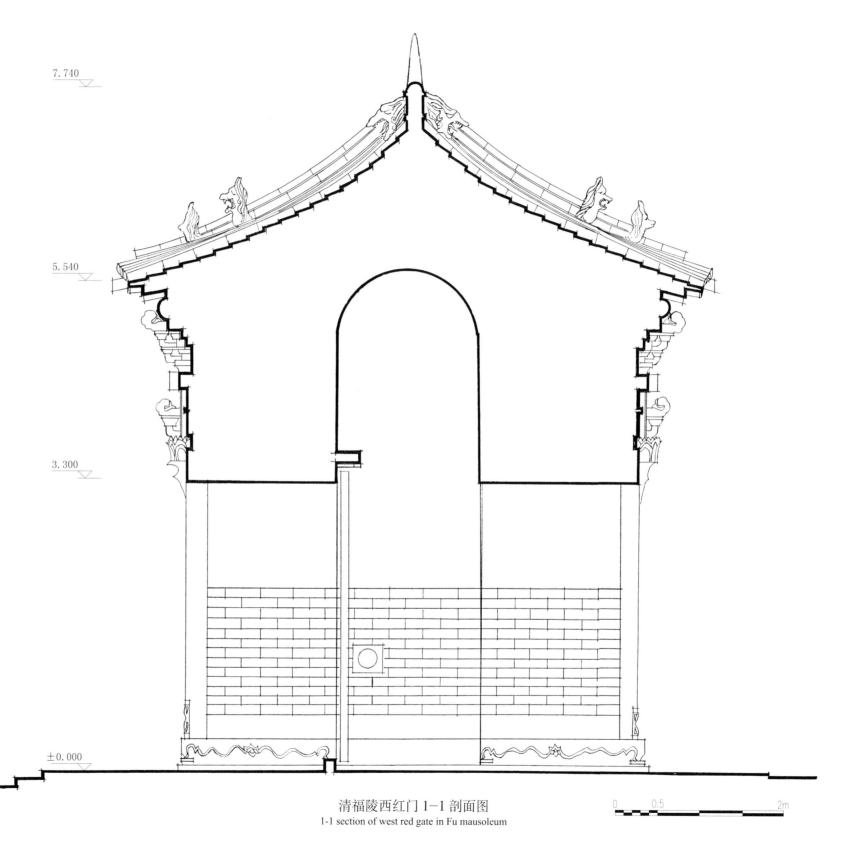

7.740

5.540

3.300

±0.000

-1.54

清福陵西红门 1-1 剖面图
1-1 section of west red gate in Fu mausoleum

0   0.5   2m

清福陵石牌楼正立面图

Elevation of stone archway in Fu mausoleum

0   1   2m

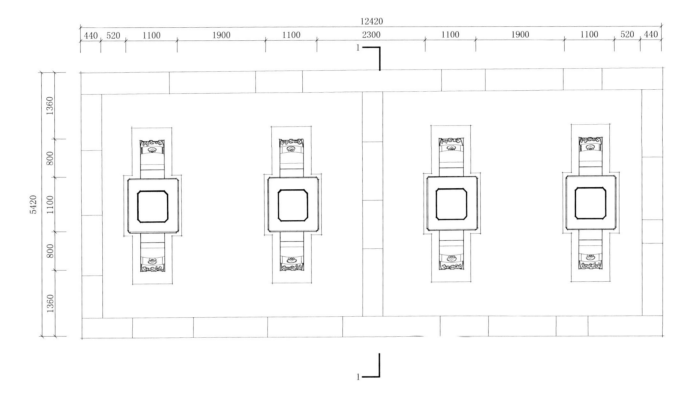

清福陵石牌楼平面图
Plan of stone archway in Fu mausoleum

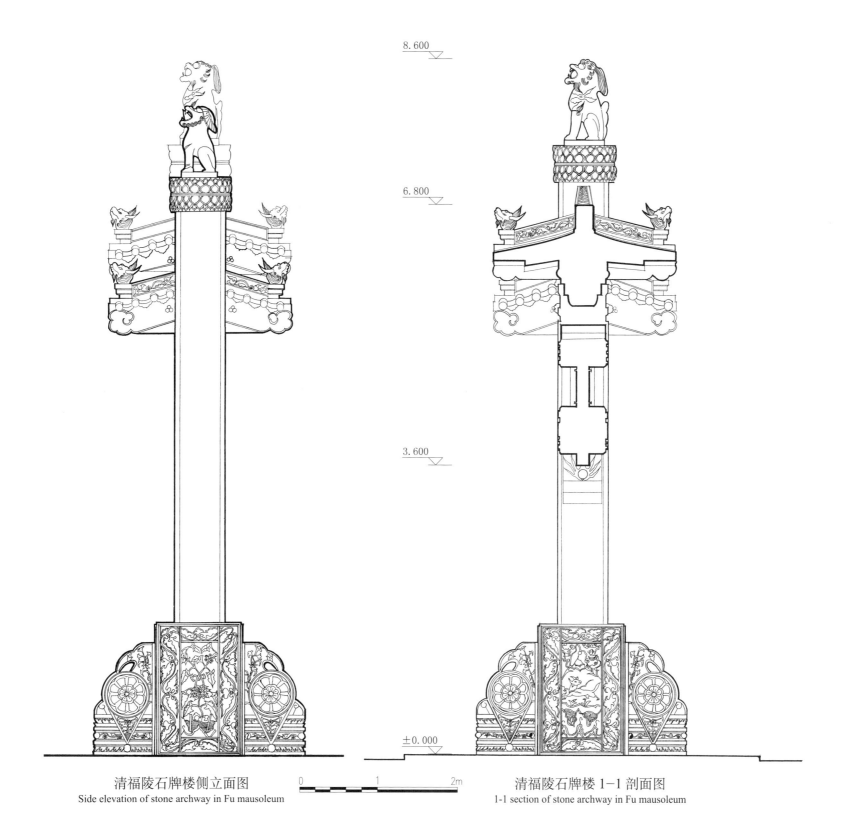

8.600

6.800

3.600

±0.000

0　　　　1　　　　2m

清福陵石牌楼侧立面图
Side elevation of stone archway in Fu mausoleum

清福陵石牌楼 1-1 剖面图
1-1 section of stone archway in Fu mausoleum

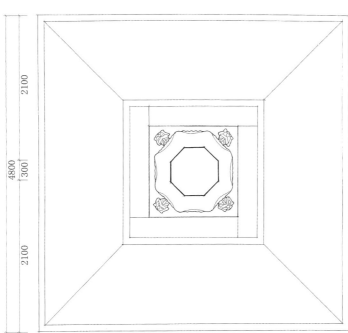

0 250 750mm

清福陵正红门内华表立面及平面图

Elevation and plan of ornamental columns inside main red gate of Fu mausoleum

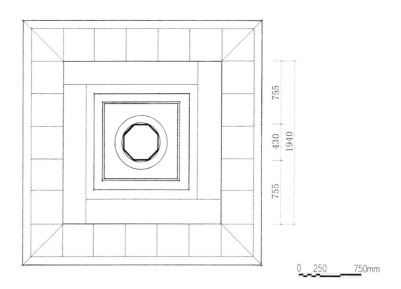

0 250 750mm

清福陵正红门外华表立面及平面图

Elevation and plan of ornamental columns outside main red gate of Fu mausoleum

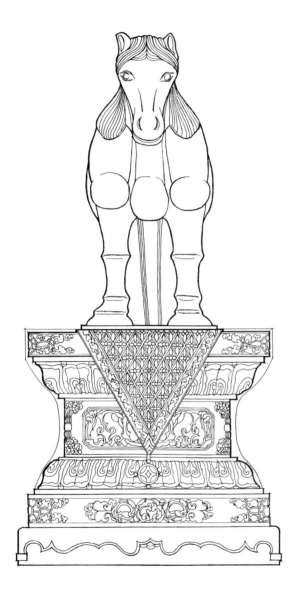

清福陵石象生立马大样图
Stone statues of Fu mausoleum-standing horse

0 10   60cm

清福陵石象生卧驼大样图
Stone statues of Fu mausoleum-crouching camel

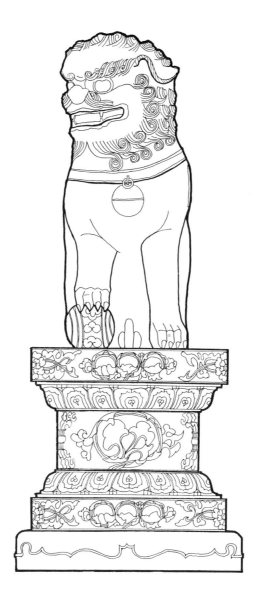

清福陵石象生坐狮大样图（一）
Stone statues of Fu mausoleum-sitting lion（I）

清福陵正红门外石狮大样图
Stone lions outside main red gate of Fu Mausoleum

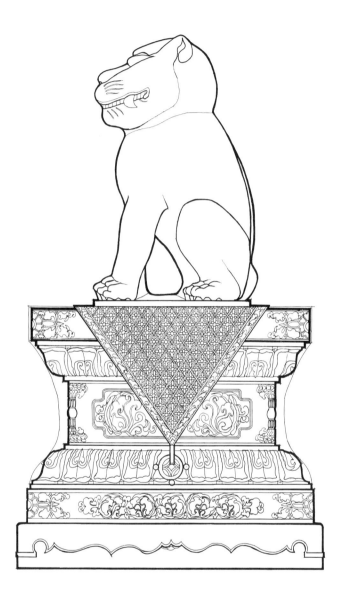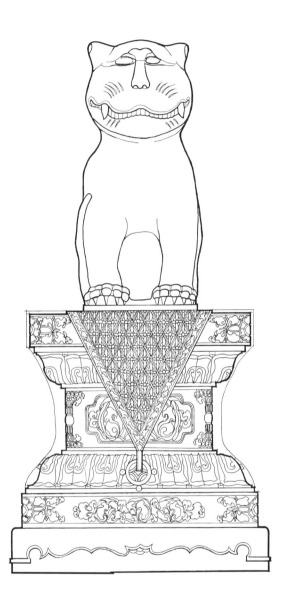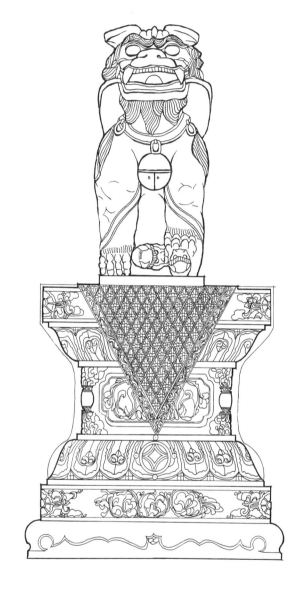

清福陵石象生坐虎大样图
Stone statues of Fu mausoleum-sitting tiger

清福陵石象生坐狮大样图（二）
Stone statues of Fu mausoleum-sitting lion (II)

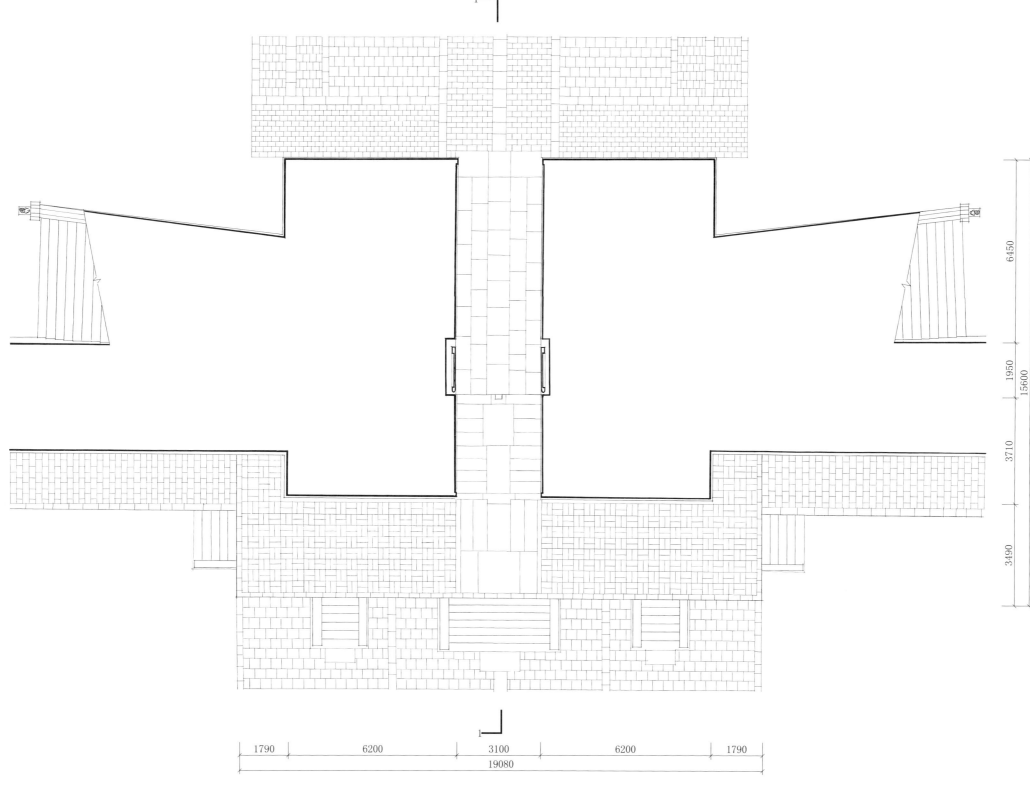

清福陵隆恩门门洞平面图
Plan of gate hole of Longen gate in Fu mausoleum

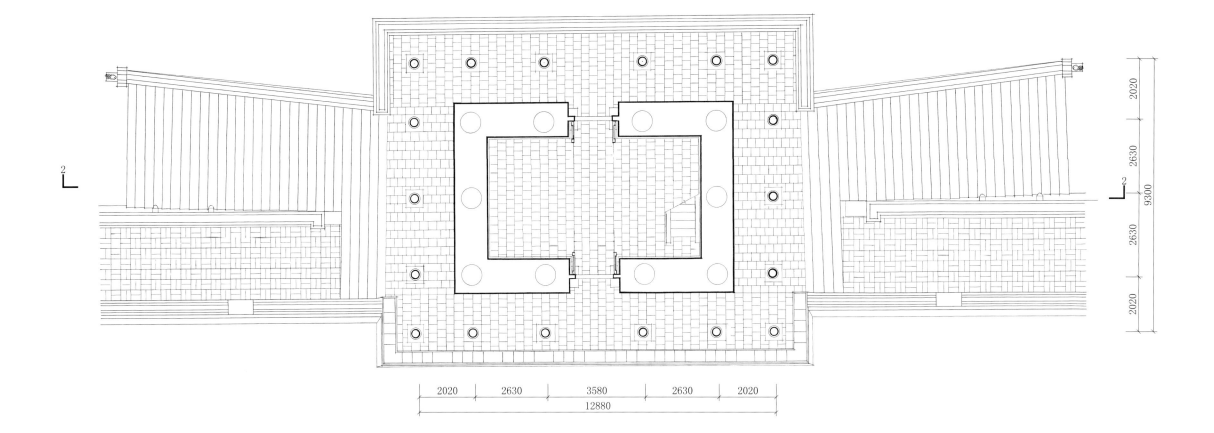

清福陵隆恩门门楼一层平面图
First floor plan of Longen gate tower in Fu mausoleum

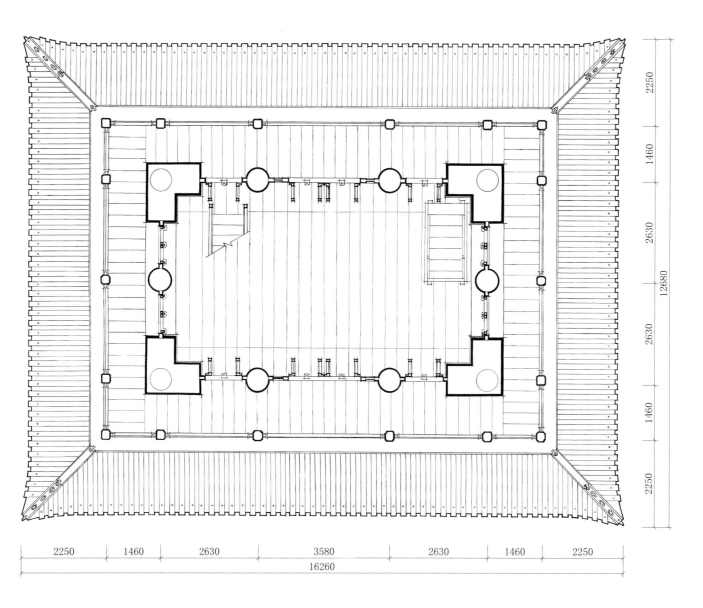

清福陵隆恩门门楼二层平面图
Second floor plan of Longen gate tower in Fu Mausoleum

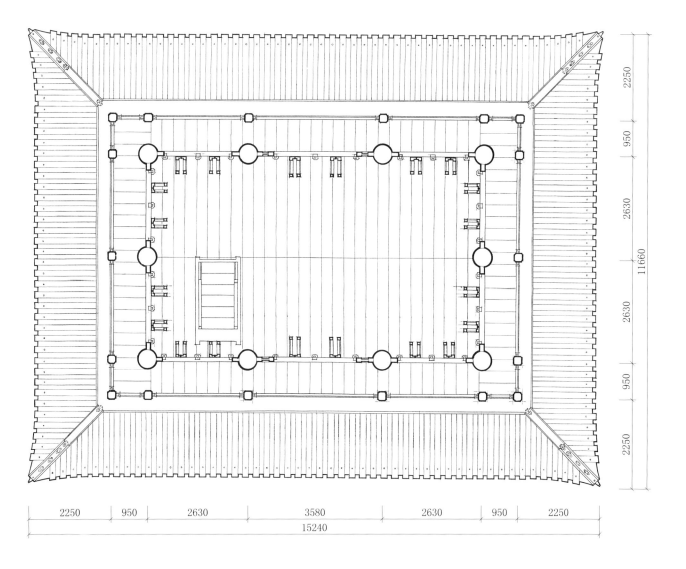

清福陵隆恩门门楼三层平面图
Third floor plan of Longen gate tower in Fu mausoleum

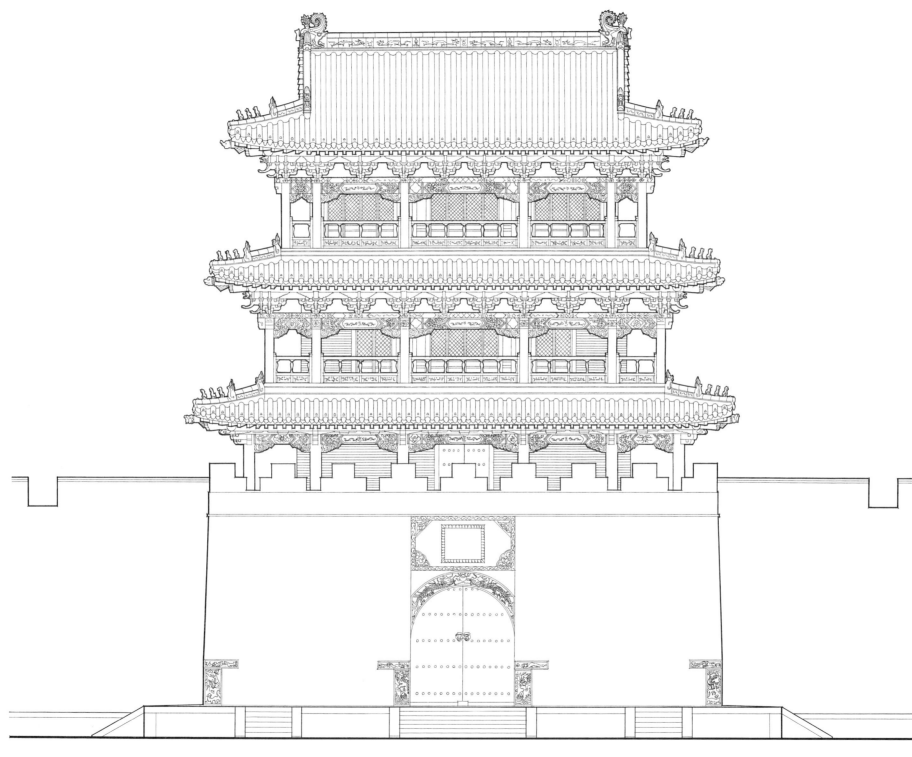

清福陵隆恩门正立面图
Elevation of Longen gate in Fu mausoleum

0 0.5   2m

清福陵隆恩门侧立面图
Side elevation of Longen gate in Fu mausoleum

0 0.5   2m

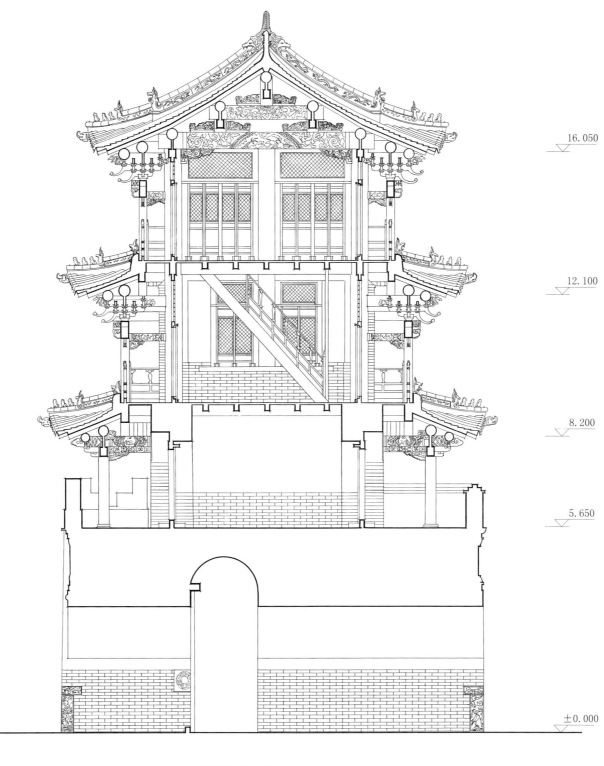

16.050

12.100

8.200

5.650

±0.000

−0.870

清福陵隆恩门 1−1 剖面图

1-1 section of Longen gate in Fu mausoleum

0 0.5 2m

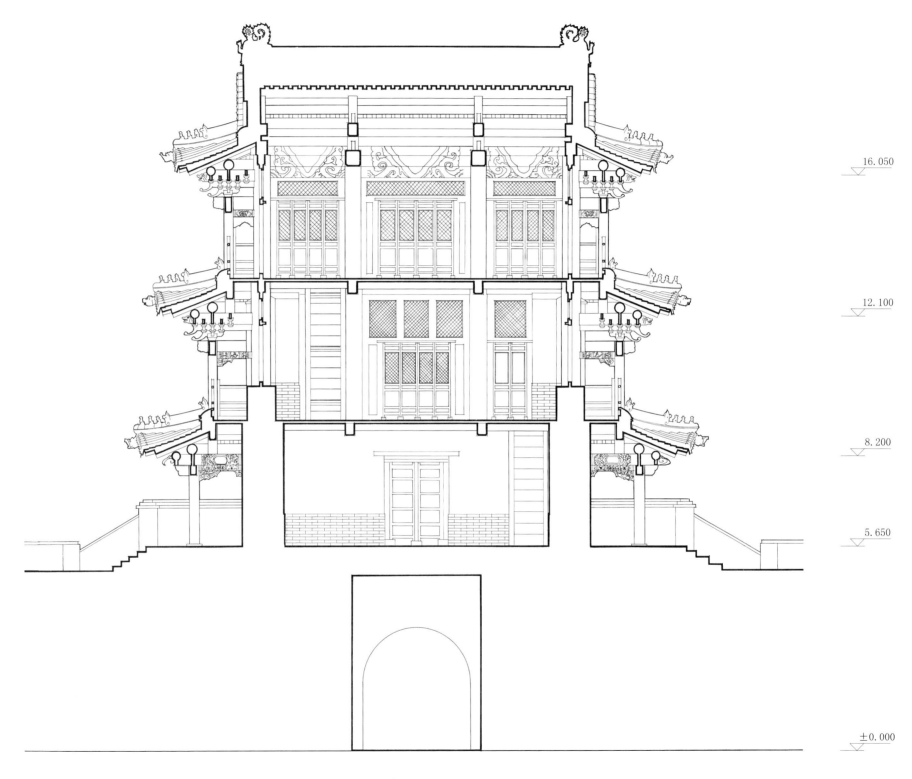

16. 050

12. 100

8. 200

5. 650

±0. 000

0 0.5   2m

清福陵隆恩门 2-2 剖面图
2-2 section of Longen gate in Fu mausoleum

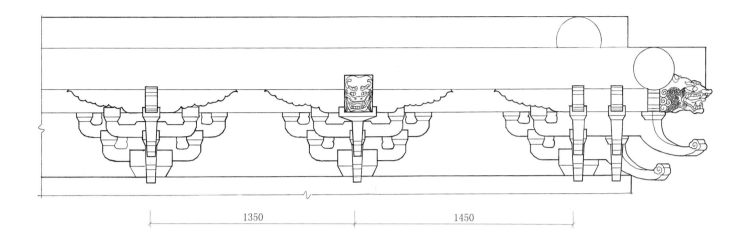

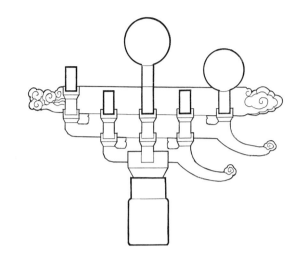

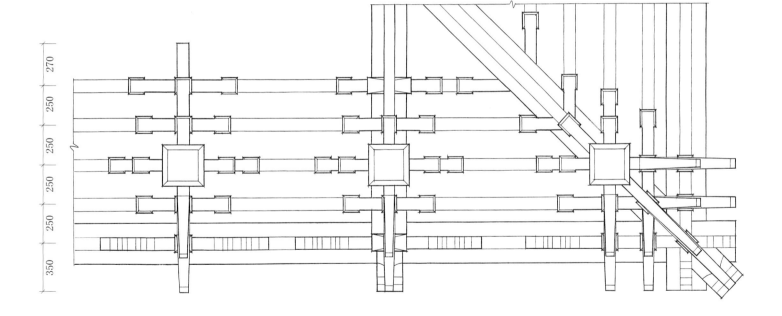

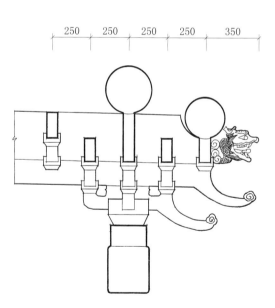

清福陵隆恩门斗栱大样图

Detail drawings of Dougong on Longen gate in Fu mausoleum

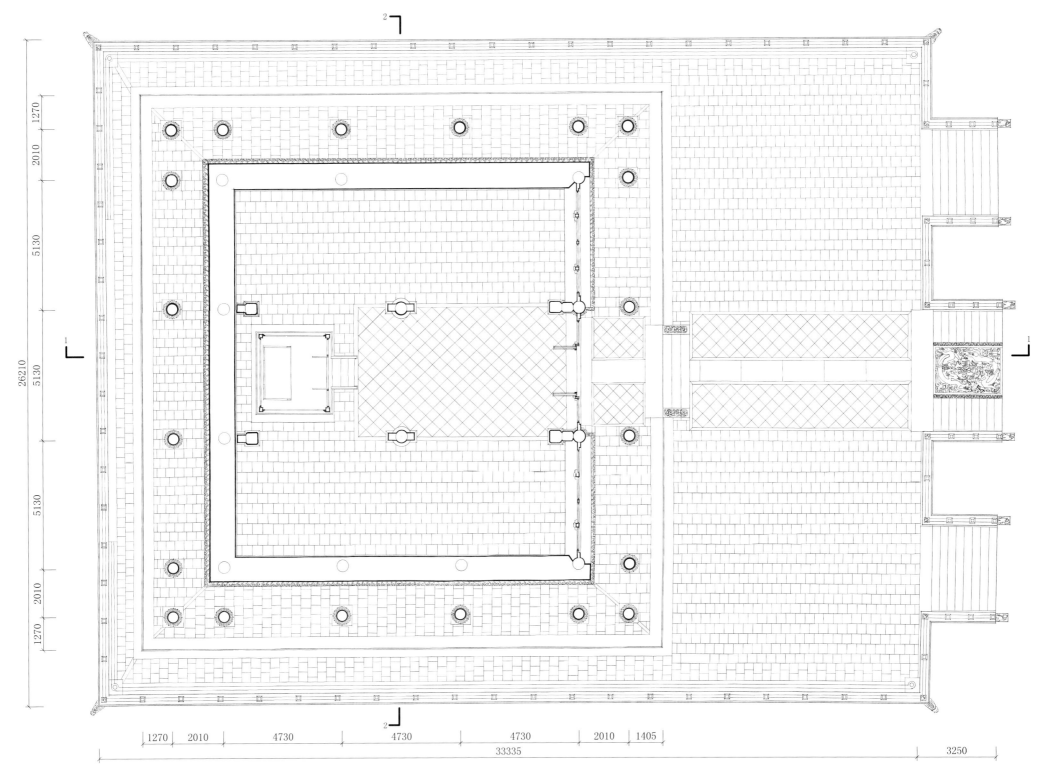

清福陵隆恩殿平面图
Plan of Longen hall in Fu mausoleum

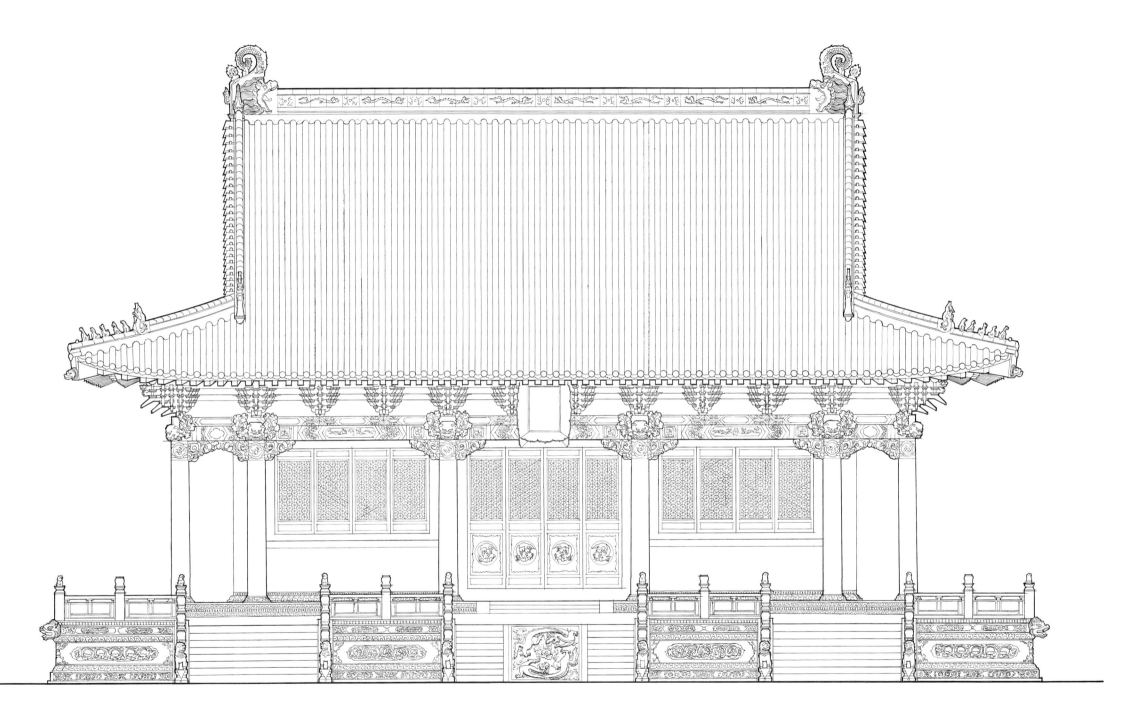

清福陵隆恩殿正立面图

Elevation of Longen hall in Fu mausoleum

0 0.5   2m

清福陵隆恩殿东立面图
East elevation of Longen hall in Fu mausoleum

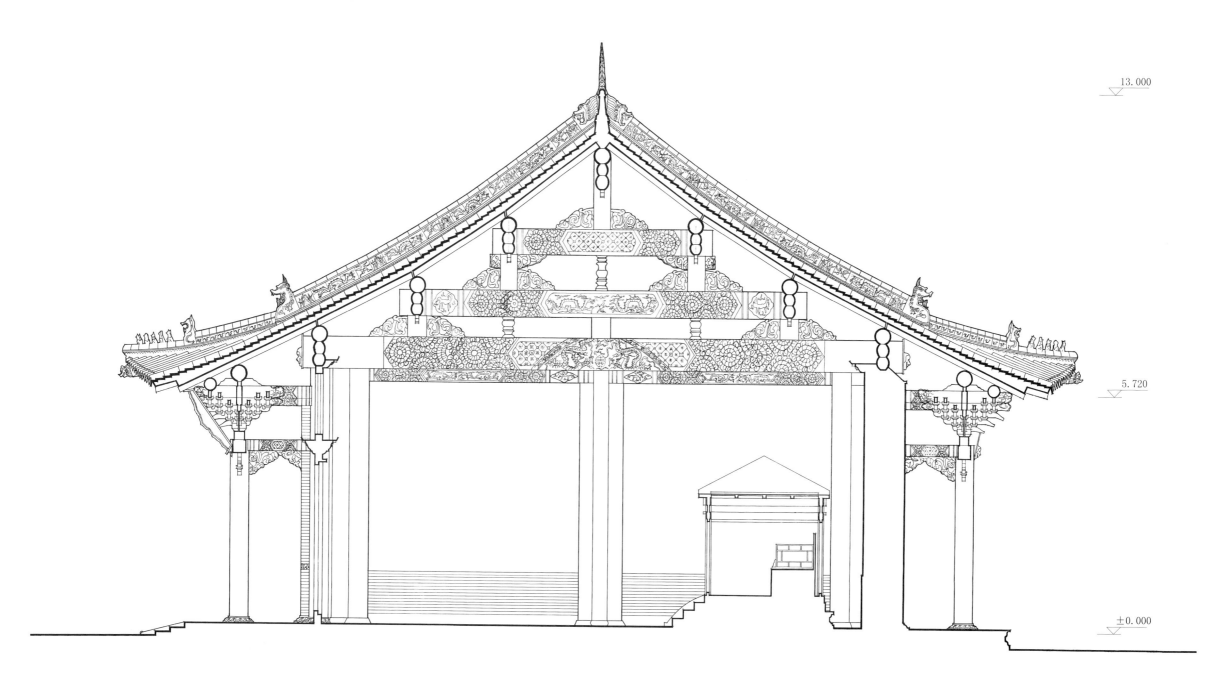

13.000

5.720

±0.000

清福陵隆恩殿 1-1 剖面图
1-1 section of Longen hall in Fu mausoleum

0 0.5   2m

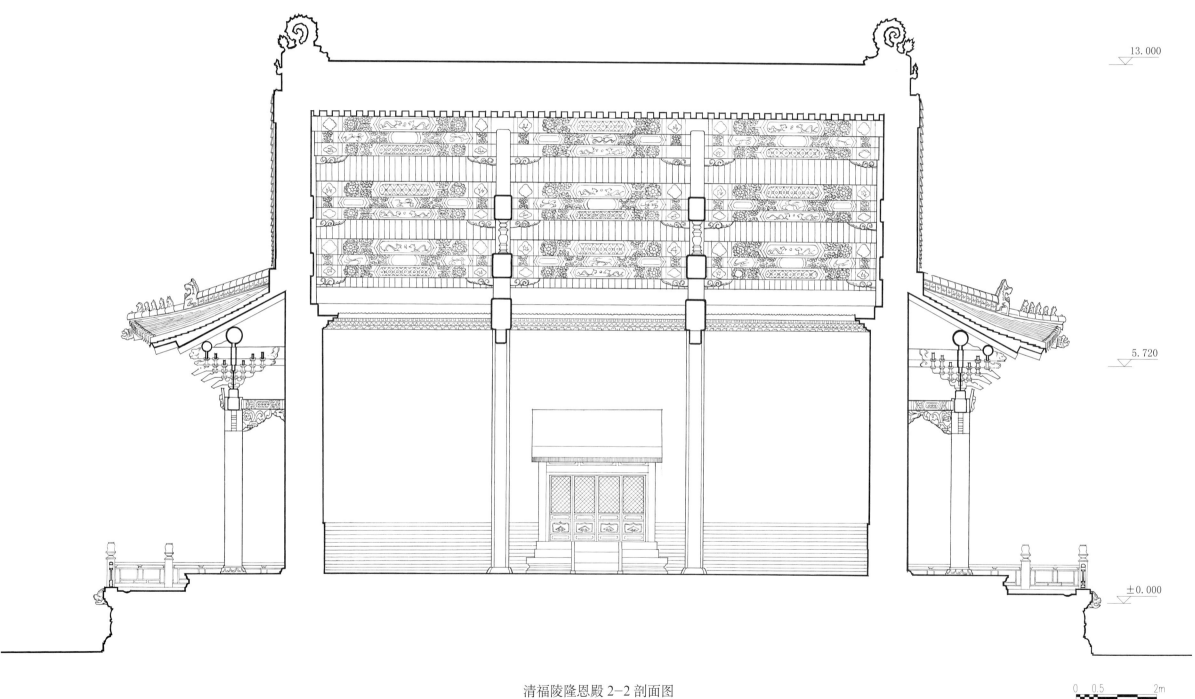

13.000

5.720

±0.000

清福陵隆恩殿 2-2 剖面图
2-2 section of Longen hall in Fu mausoleum

0 0.5 2m

清福陵隆恩殿台基大样图

Detail drawing of stylobate on Longen hall in Fu mausoleum

清福陵隆恩殿云龙石大样图
Detail drawing of cloud dragon stone on Longen hall in Fu mausoleum

0   0.1   0.5m

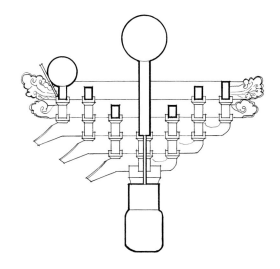
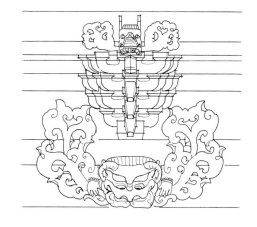
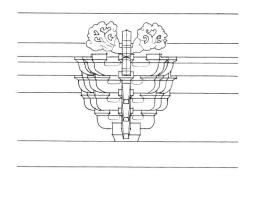
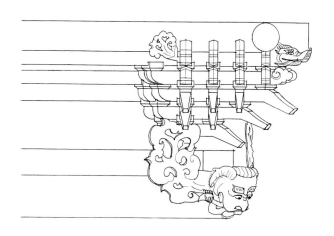

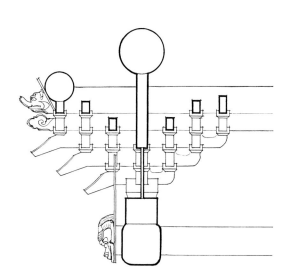
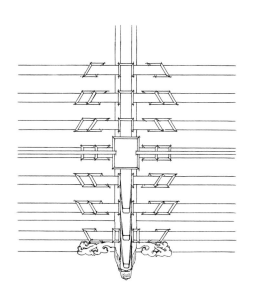
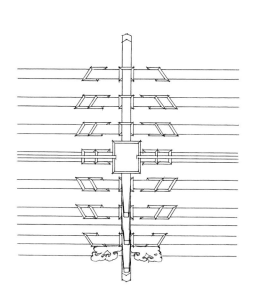
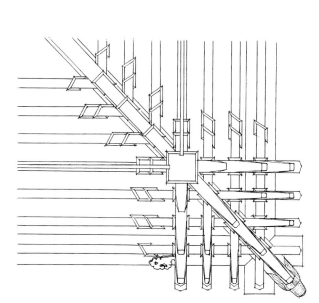

0   0.25   0.75m

清福陵隆恩殿斗栱大样图

Detail drawings of Dougong on Longen hall in Fu mausoleum

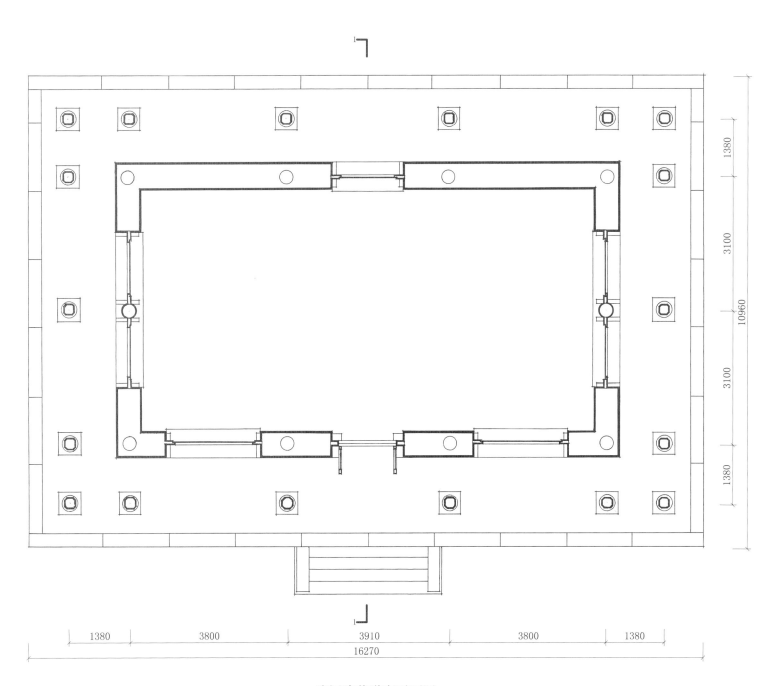

清福陵茶膳房平面图
Plan of kitchen and washhouse in Fu mausoleum

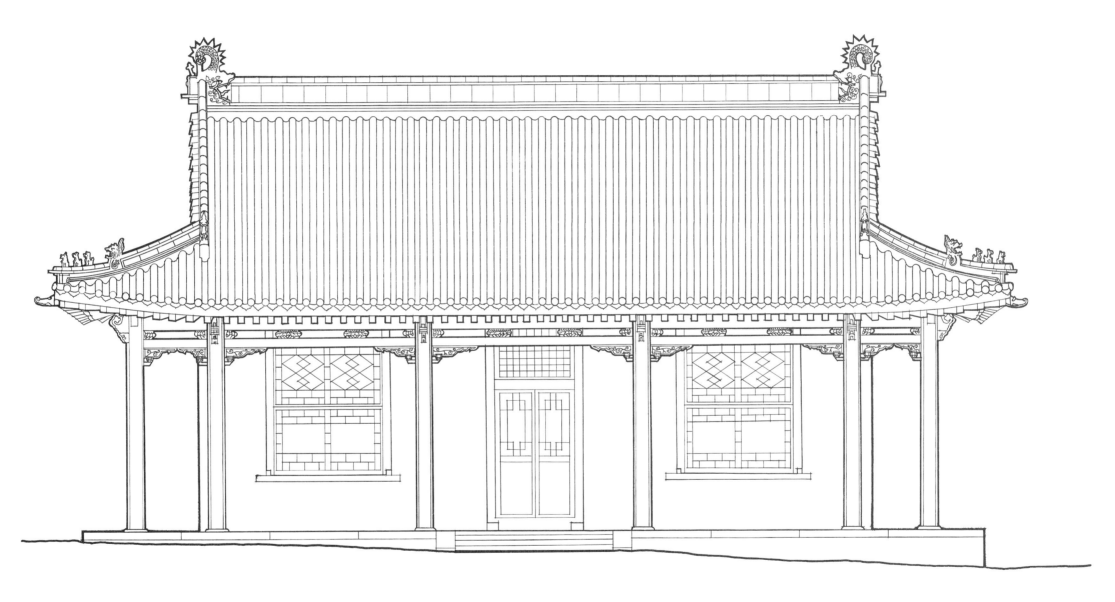

0   0.5   2m

清福陵茶膳房正立面图
Elevation of kitchen and washhouse in Fu mausoleum

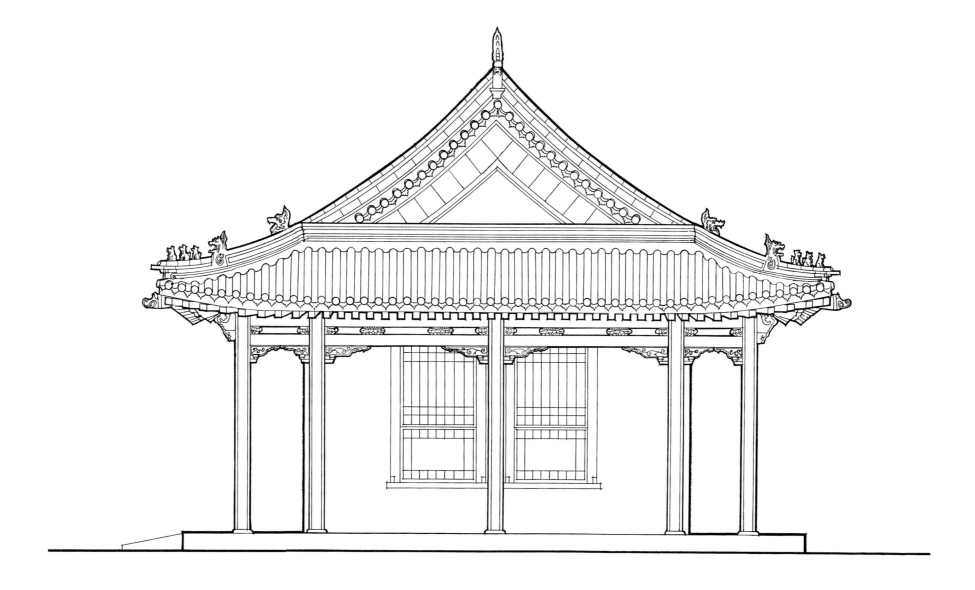

清福陵茶膳房侧立面图
Side elevation of kitchen and washhouse in Fu mausoleum

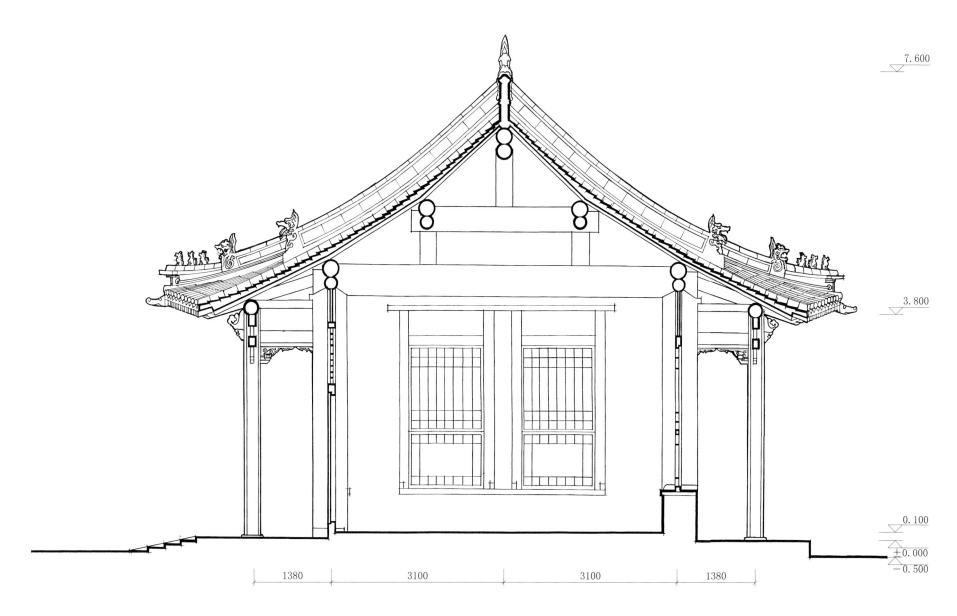

7.600

3.800

0.100

±0.000

−0.500

1380　3100　3100　1380

清福陵茶膳房 1−1 剖面图
1-1 section of kitchen and washhouse in Fu mausoleum

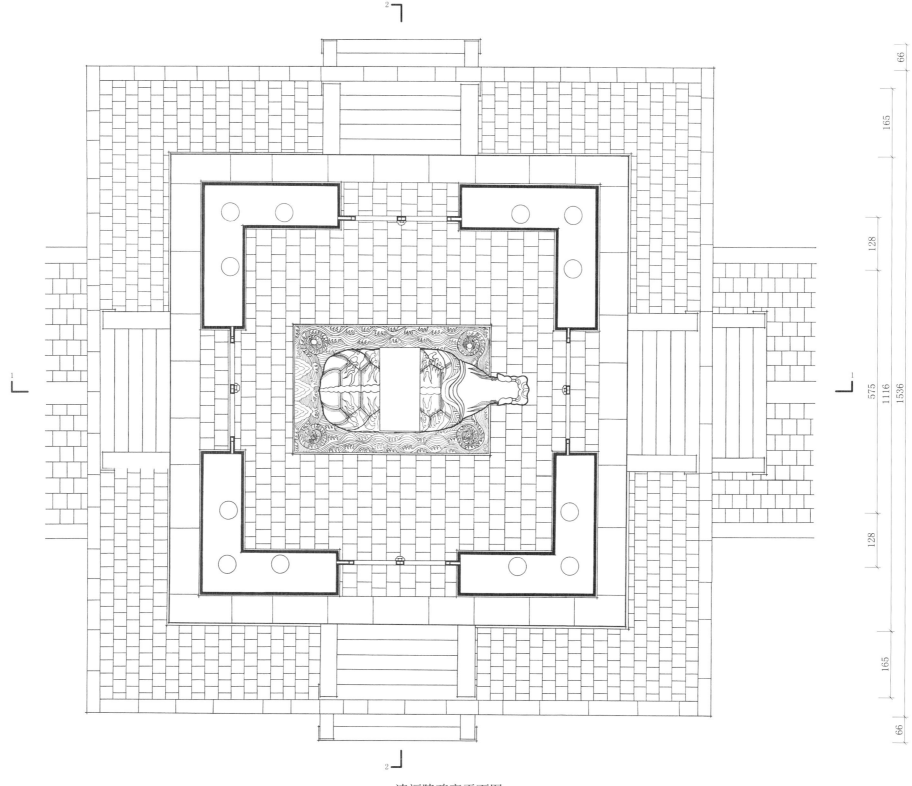

清福陵碑亭平面图
Plan of stele pavilion in Fu mausoleum

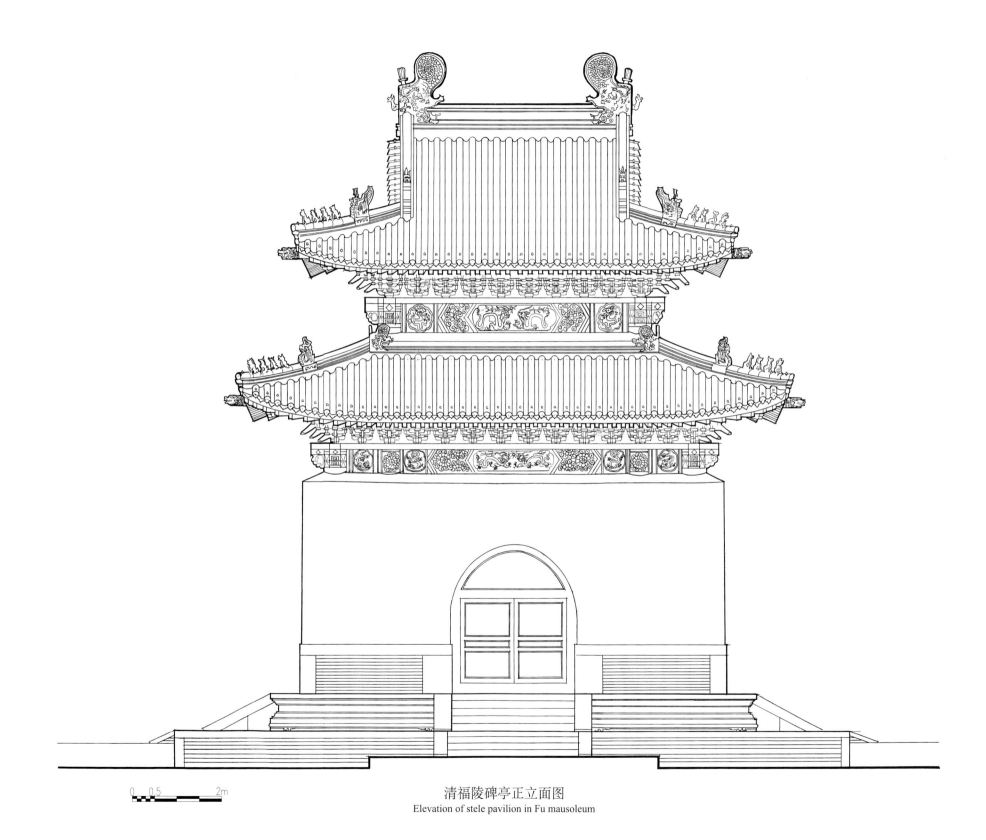

清福陵碑亭正立面图
Elevation of stele pavilion in Fu mausoleum

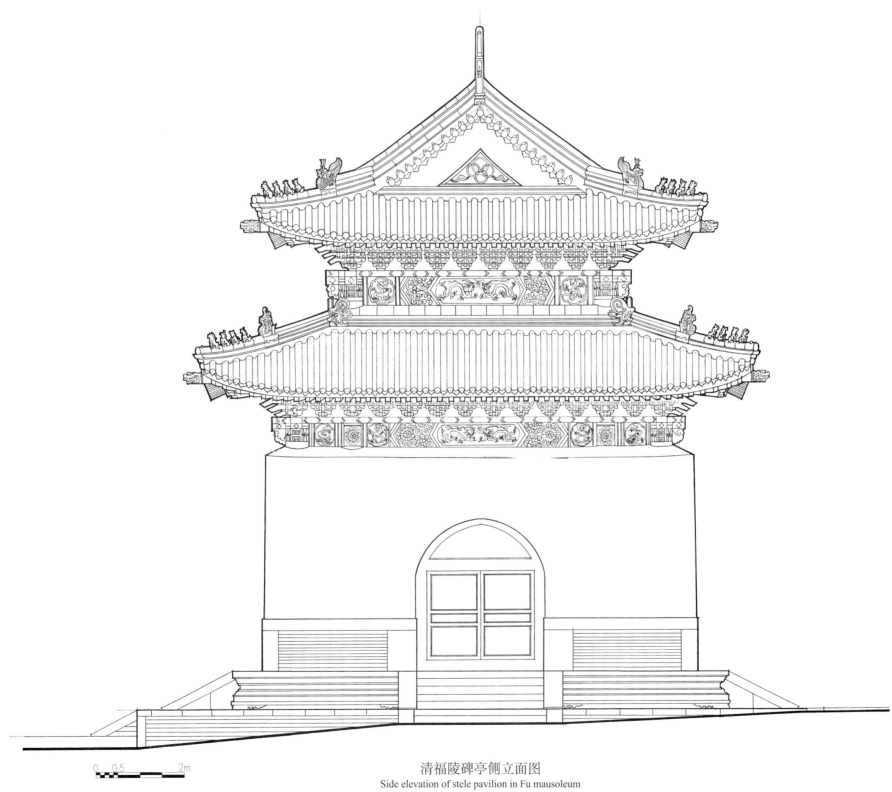

0　0.5　　　2m

清福陵碑亭侧立面图
Side elevation of stele pavilion in Fu mausoleum

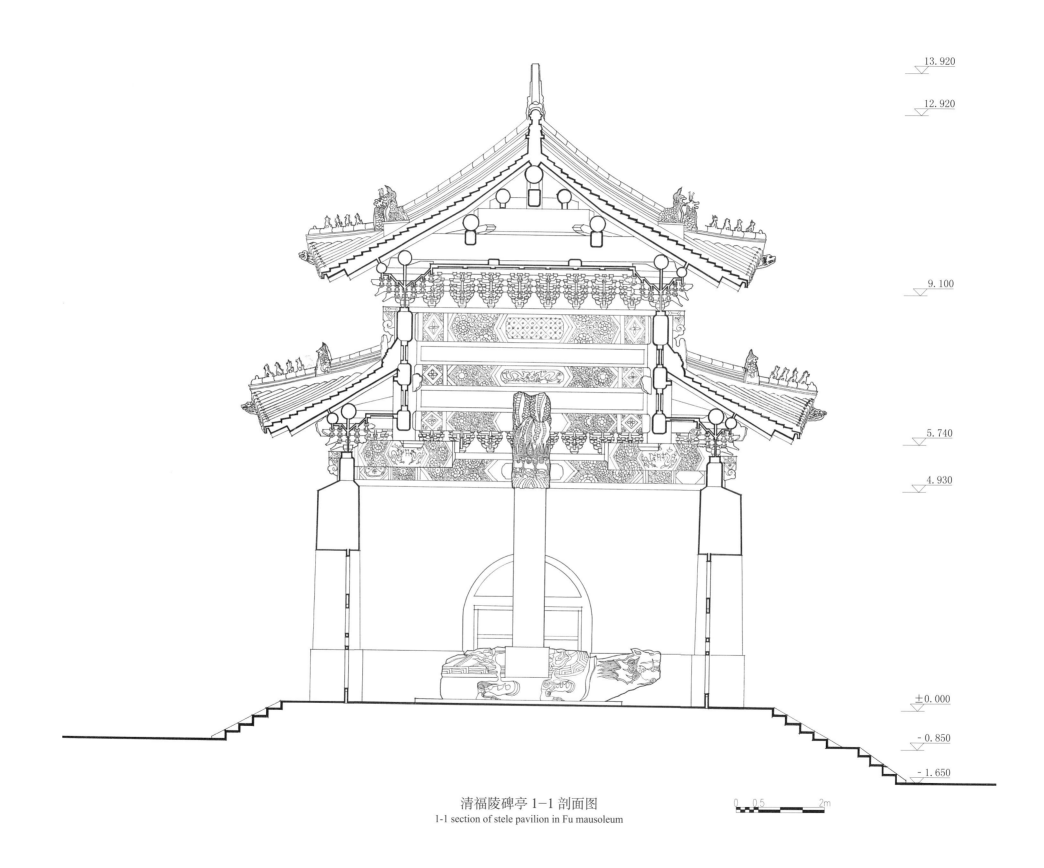

13.920

12.920

9.100

5.740

4.930

±0.000

-0.850

-1.650

清福陵碑亭 1-1 剖面图
1-1 section of stele pavilion in Fu mausoleum

0 0.5   2m

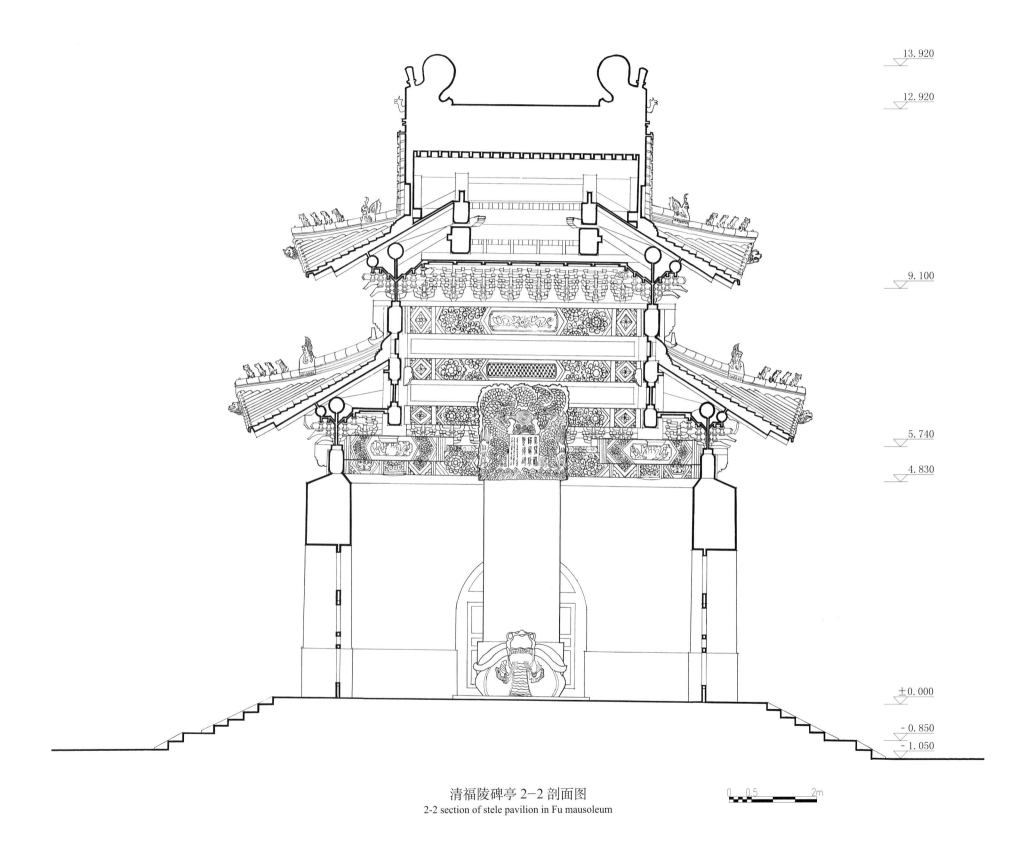

13.920

12.920

9.100

5.740

4.830

±0.000

- 0.850

- 1.050

清福陵碑亭 2—2 剖面图

2-2 section of stele pavilion in Fu mausoleum

0 0.5   2m

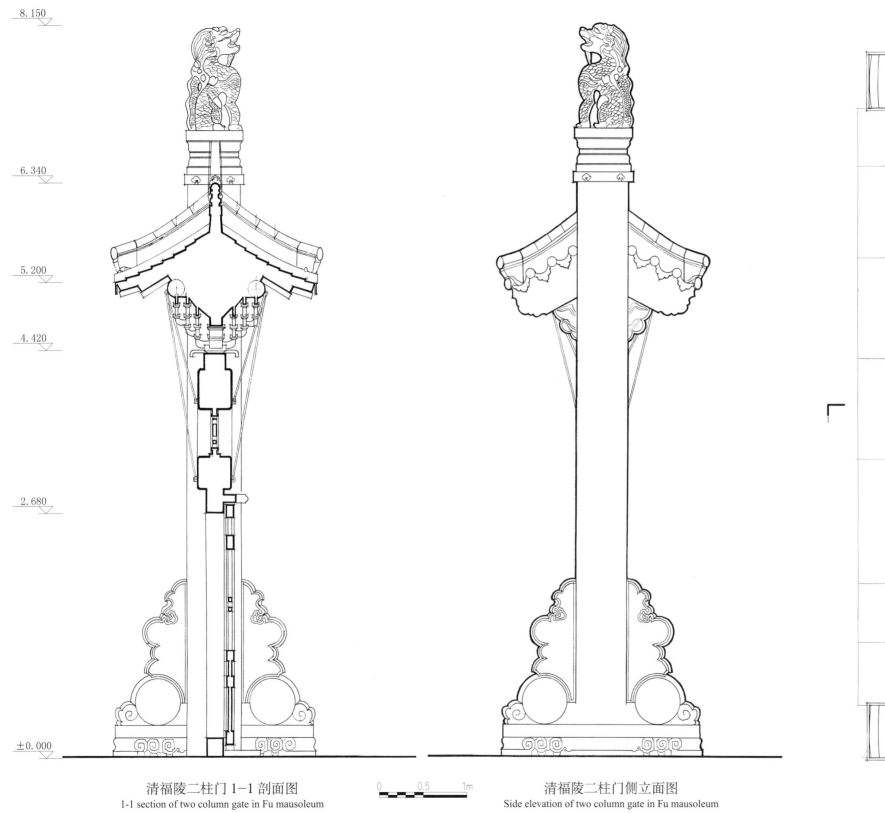

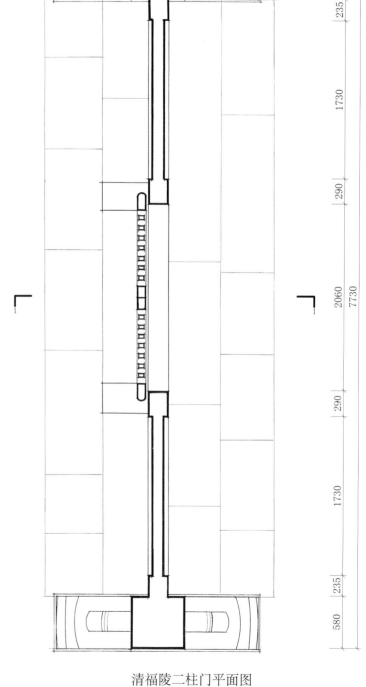

8.150

6.340

5.200

4.420

2.680

±0.000

清福陵二柱门 1-1 剖面图
1-1 section of two column gate in Fu mausoleum

0   0.5   1m

清福陵二柱门侧立面图
Side elevation of two column gate in Fu mausoleum

清福陵二柱门平面图
Plan of two column gate in Fu mausoleum

580

235

1730

290

2060   7730

290

1730

235

580

0   0.5   1m

清福陵二柱门正立面图
Elevation of two column gate in Fu mausoleum

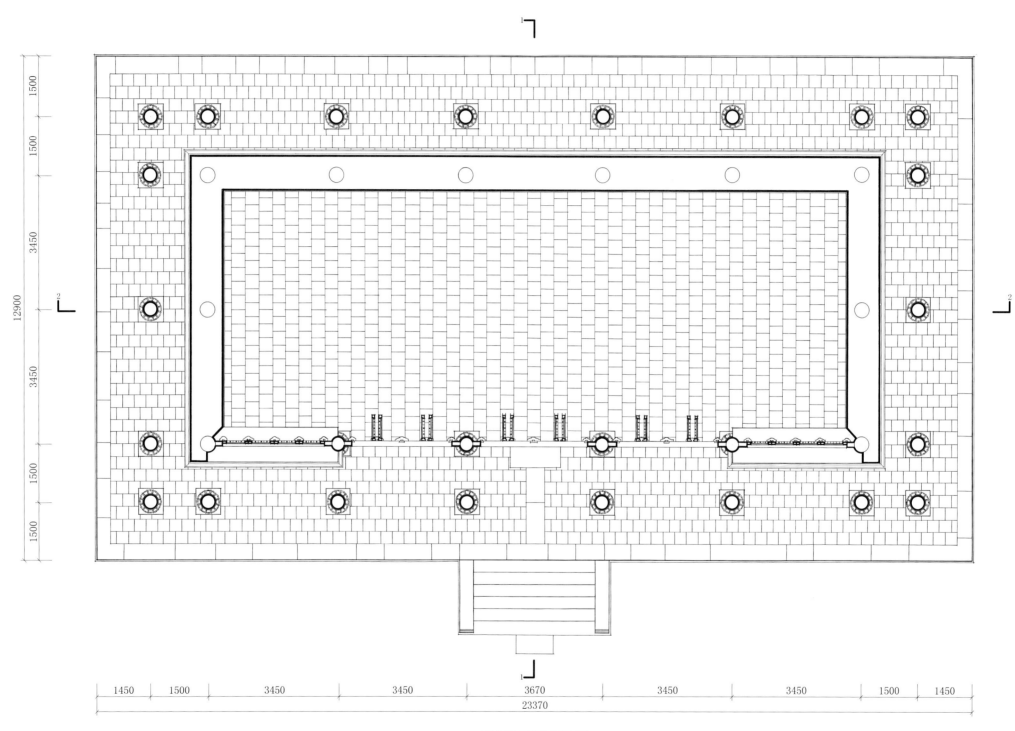

清福陵配殿平面图
Plan of supporting hall in Fu mausoleum

清福陵配殿正立面图
Elevation of supporting hall in Fu mausoleum

0　0.5　2m

清福陵配殿侧立面图
Side elevation of supporting hall in Fu mausoleum

9.680

4.600

0.080

±0.000

−0.880

清福陵配殿 1-1 剖面图
1-1 section of supporting hall in Fu mausoleum

0 0.5   2m

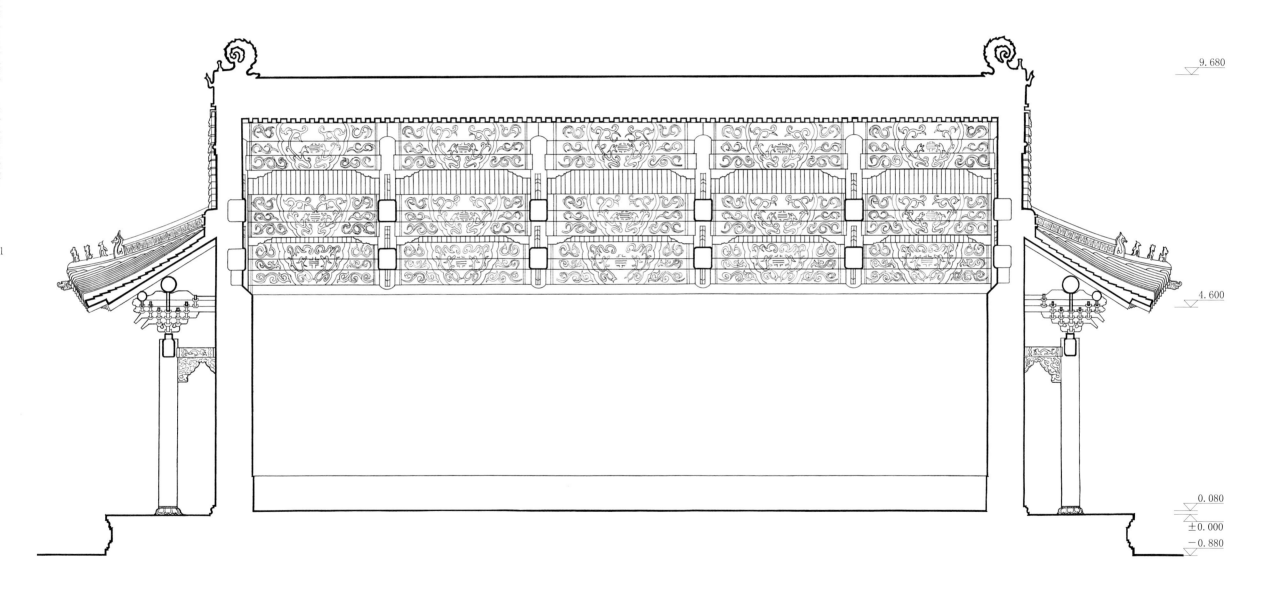

9.680

4.600

0.080

±0.000

-0.880

0 0.5   2m

清福陵配殿 2-2 剖面图

2-2 section of supporting hall in Fu mausoleum

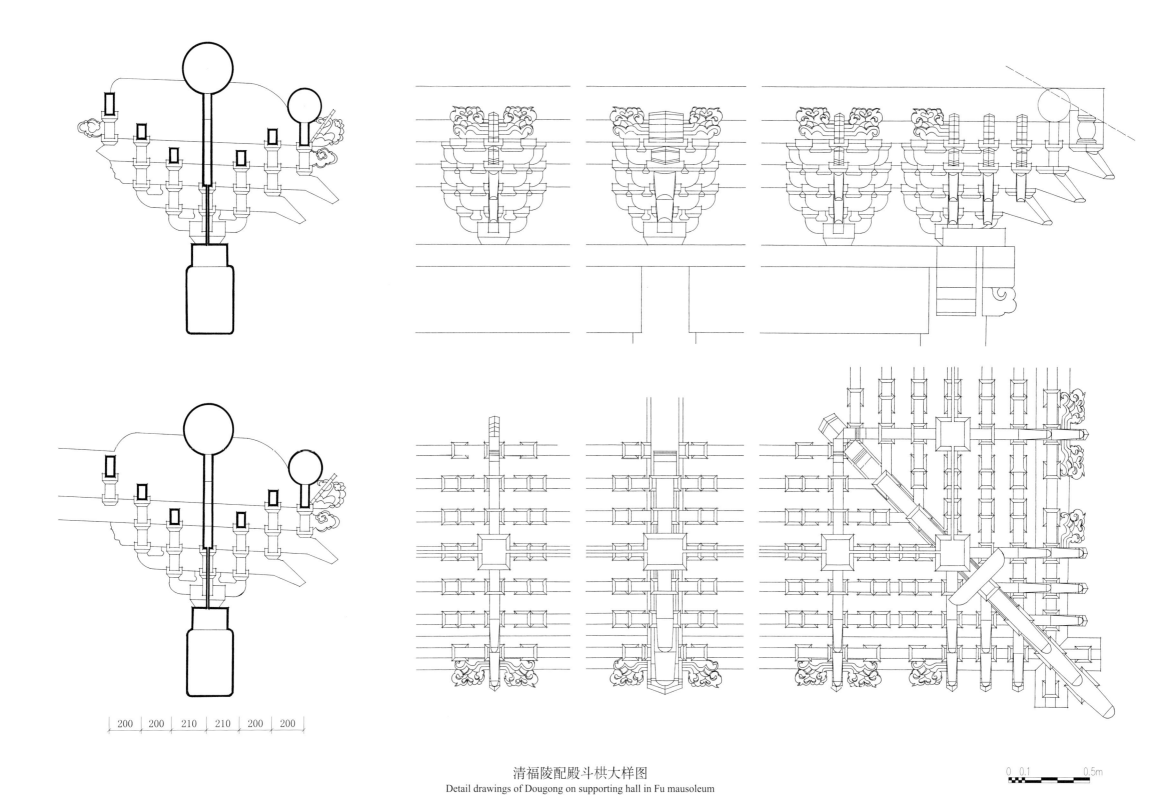

| 200 | 200 | 210 | 210 | 200 | 200 |

清福陵配殿斗栱大样图

Detail drawings of Dougong on supporting hall in Fu mausoleum

0 0.1 0.5m

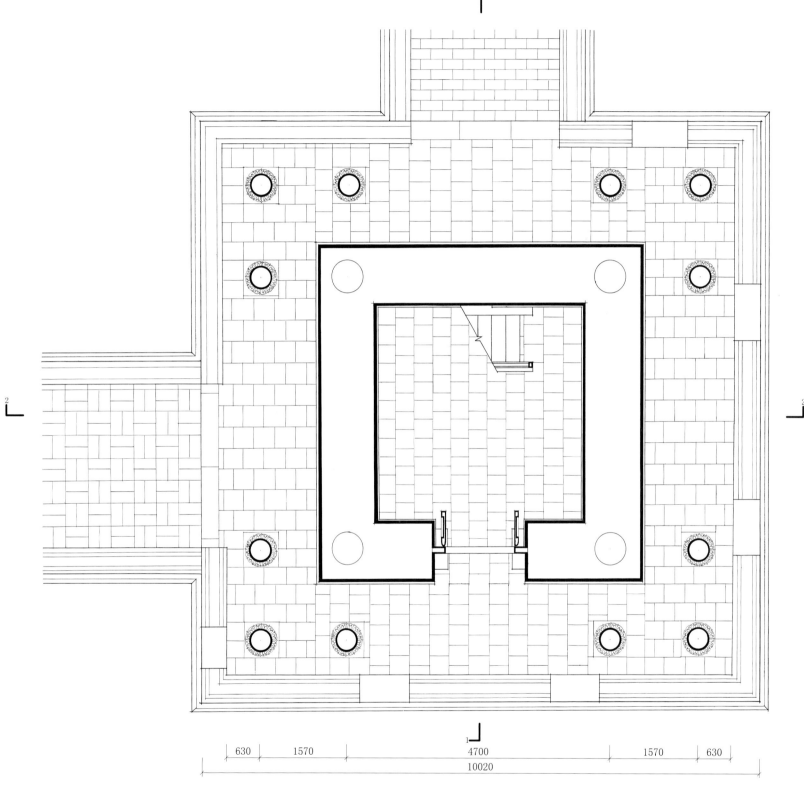

| 630 | 1570 | 4700 | 1570 | 630 |

10020

清福陵东南角楼一层平面图
First floor plan of south east corner tower in Fu mausoleum

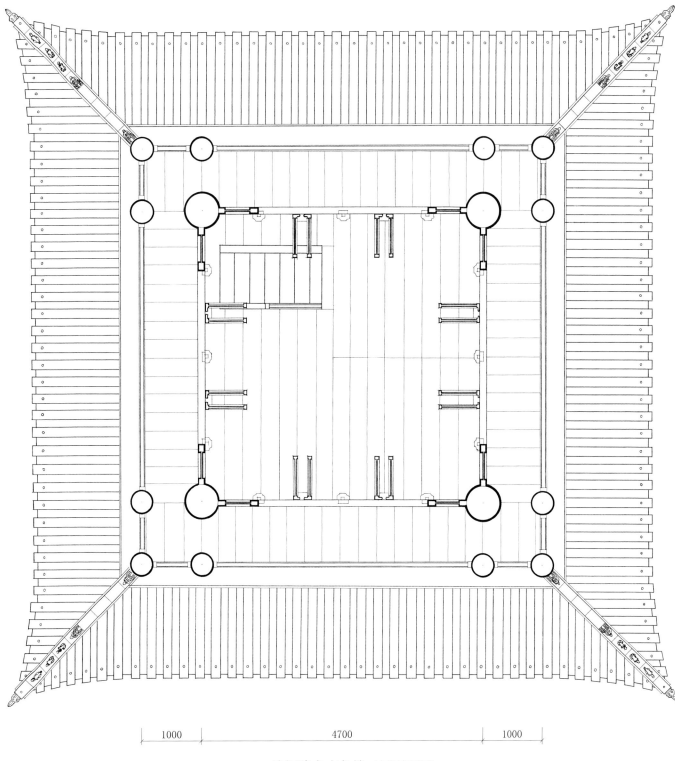

| 1000 | 4700 | 1000 |

清福陵东南角楼二层平面图
Second floor plan of south east corner tower in Fu mausoleum

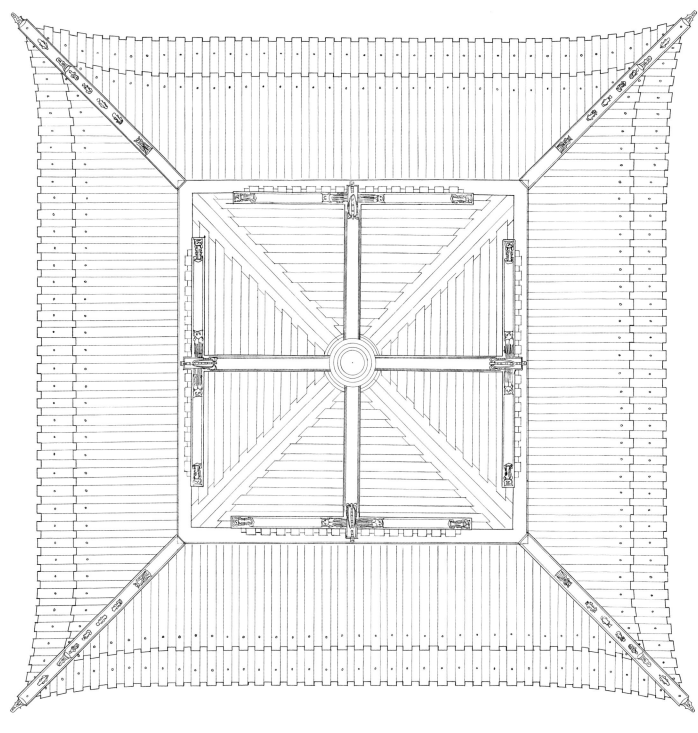

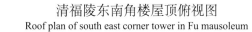

清福陵东南角楼屋顶俯视图
Roof plan of south east corner tower in Fu mausoleum

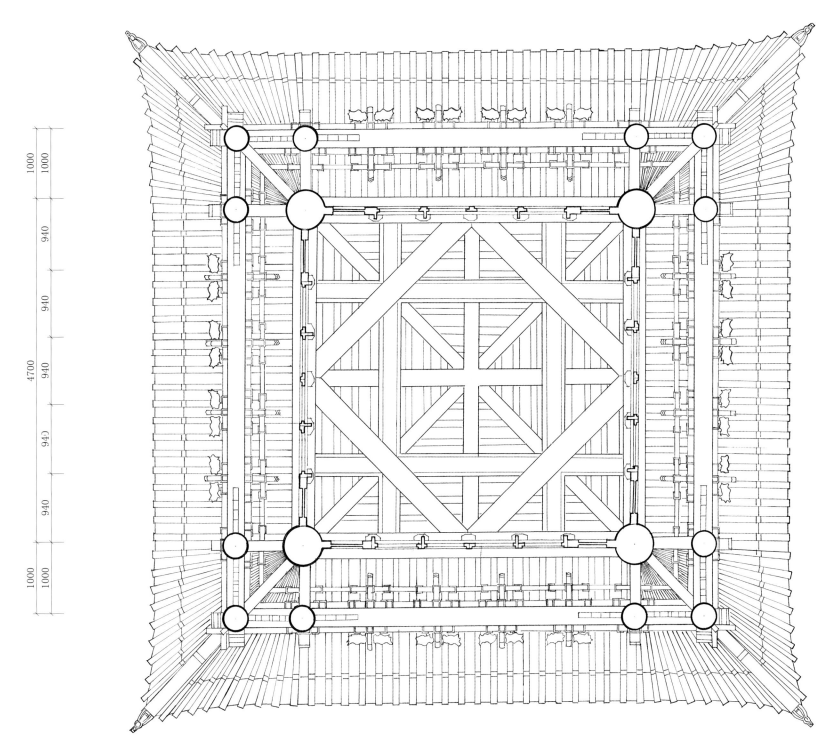

清福陵东南角楼梁架仰视图

Beam frame upward plan of south east corner tower in Fu mausoleum

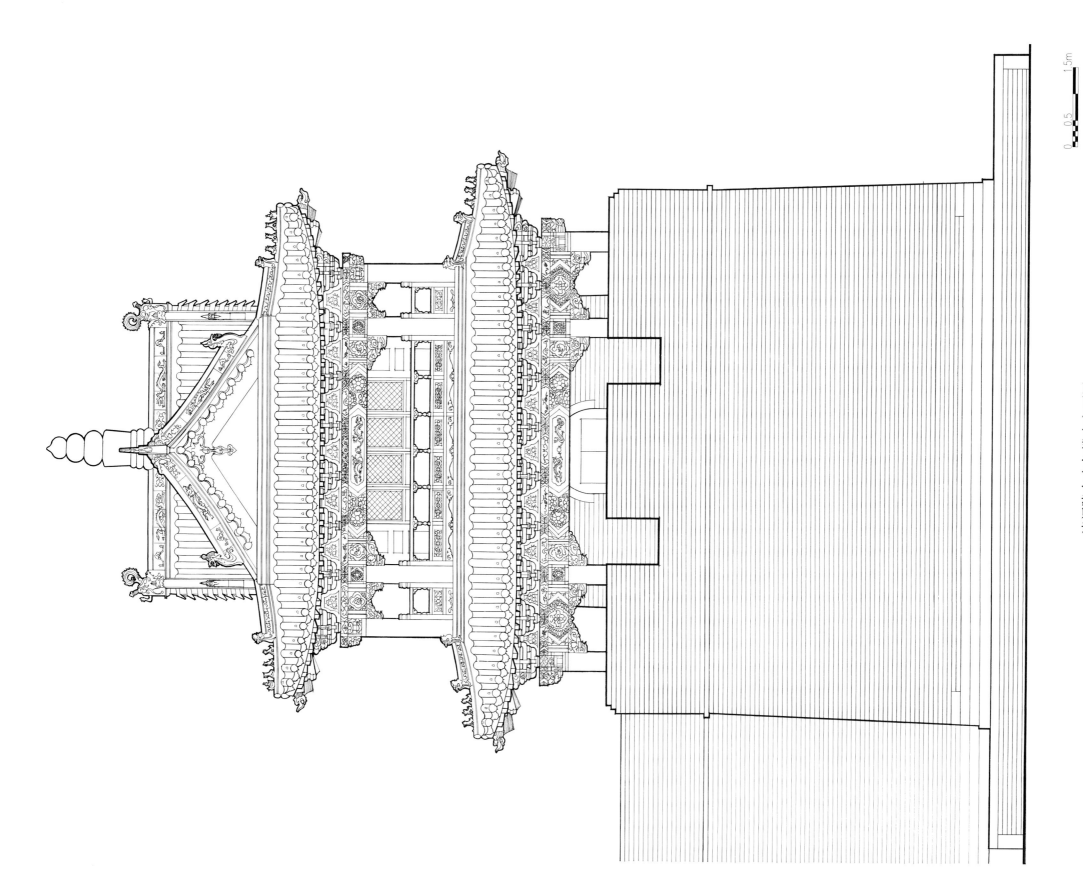

清福陵东南角楼南立面图

South elevation of south east corner tower in Fu mausoleum

0 0.5 1.5m

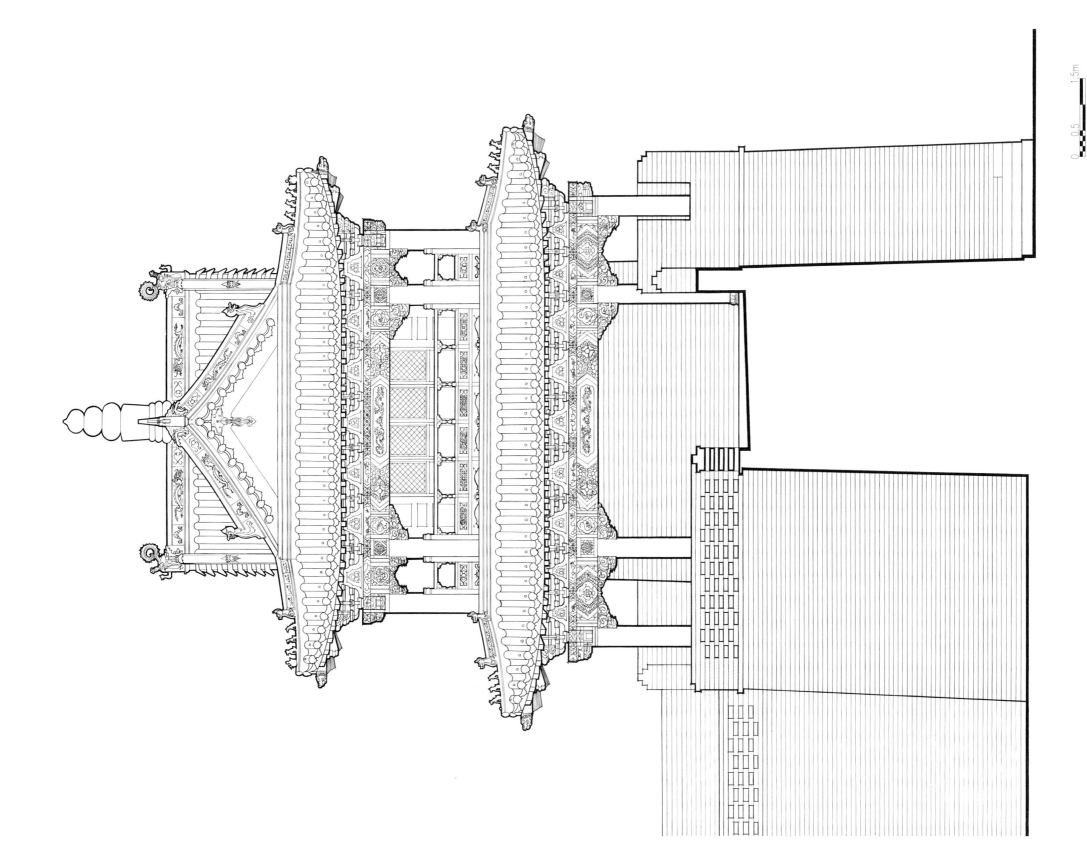

清福陵东南角角楼西立面图
West elevation of south east corner tower in Fu mausoleum

0 0.5 1.5m

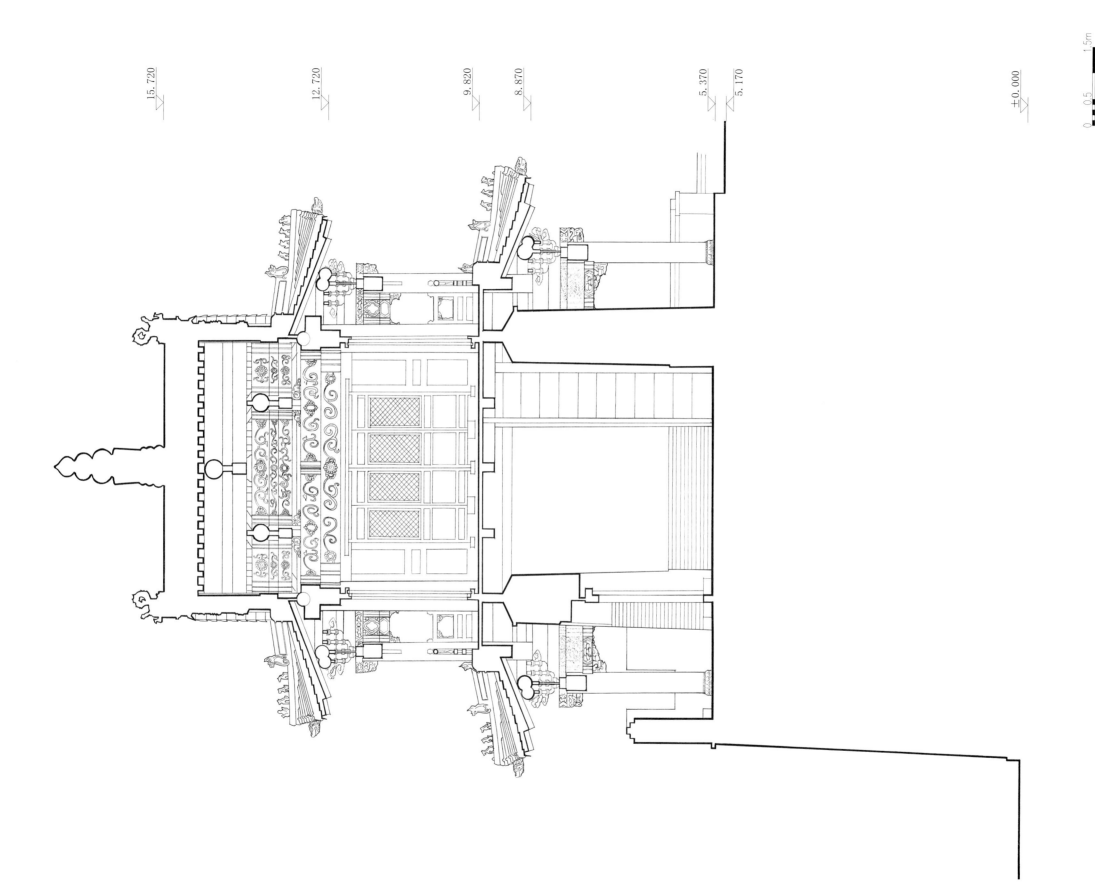

15.720

12.720

9.820

8.870

5.370
5.170

±0.000

0 0.5 1.5m

清福陵东南角楼 1-1 剖面图
1-1 section of south east corner tower in Fu mausoleum

15.720

12.720

9.820

8.870

5.370

5.170

±0.000

清福陵东南角楼 2-2 剖面图
2-2 section of south east corner tower in Fu mausoleum

0 0.5 1 5m

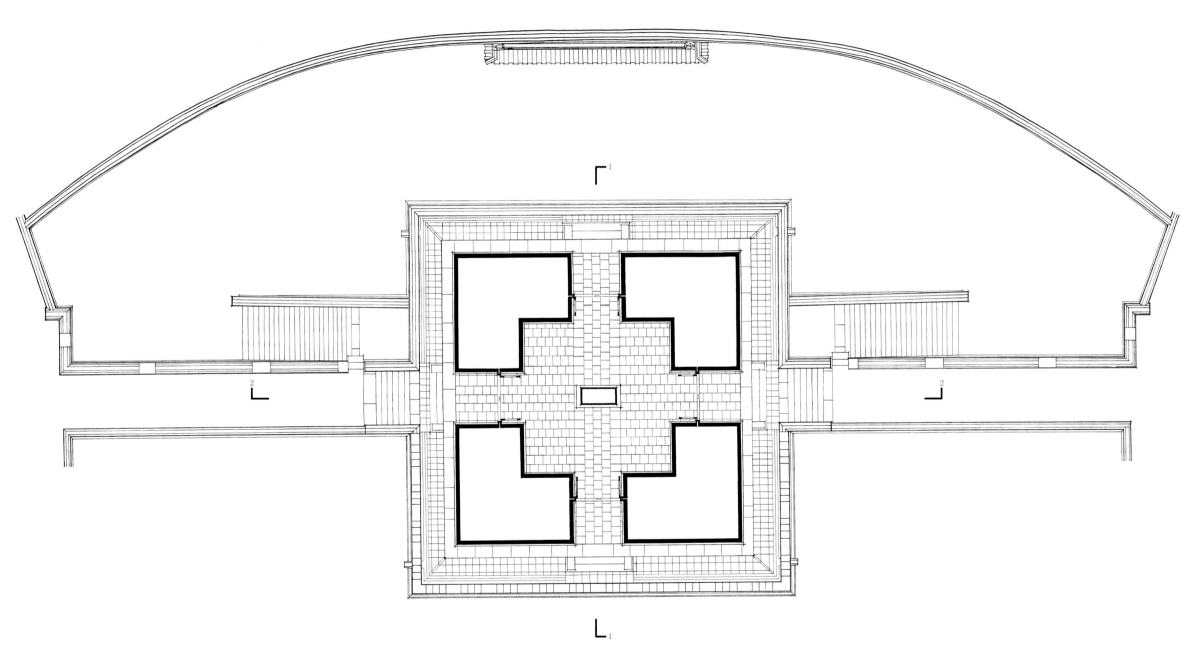

0　1　　　4m

清福陵明楼平面图
Plan of Ming tower in Fu mausoleum

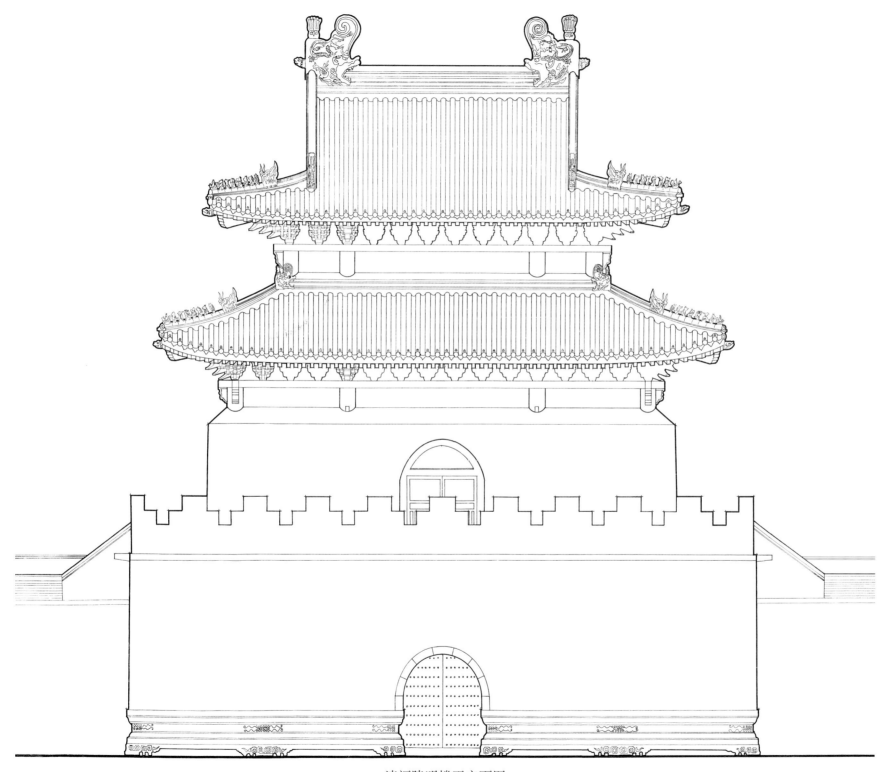

清福陵明楼正立面图
Elevation of Ming tower in Fu mausoleum

0 0.5 2m

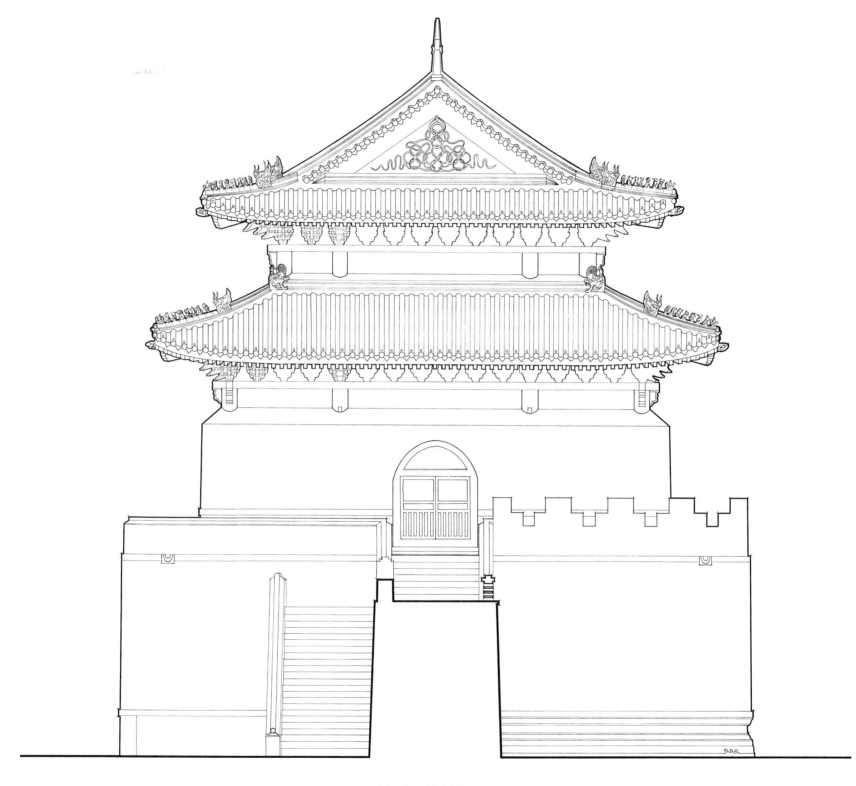

清福陵明楼侧立面图
Side elevation of Ming tower in Fu mausoleum

0 0.5   2m

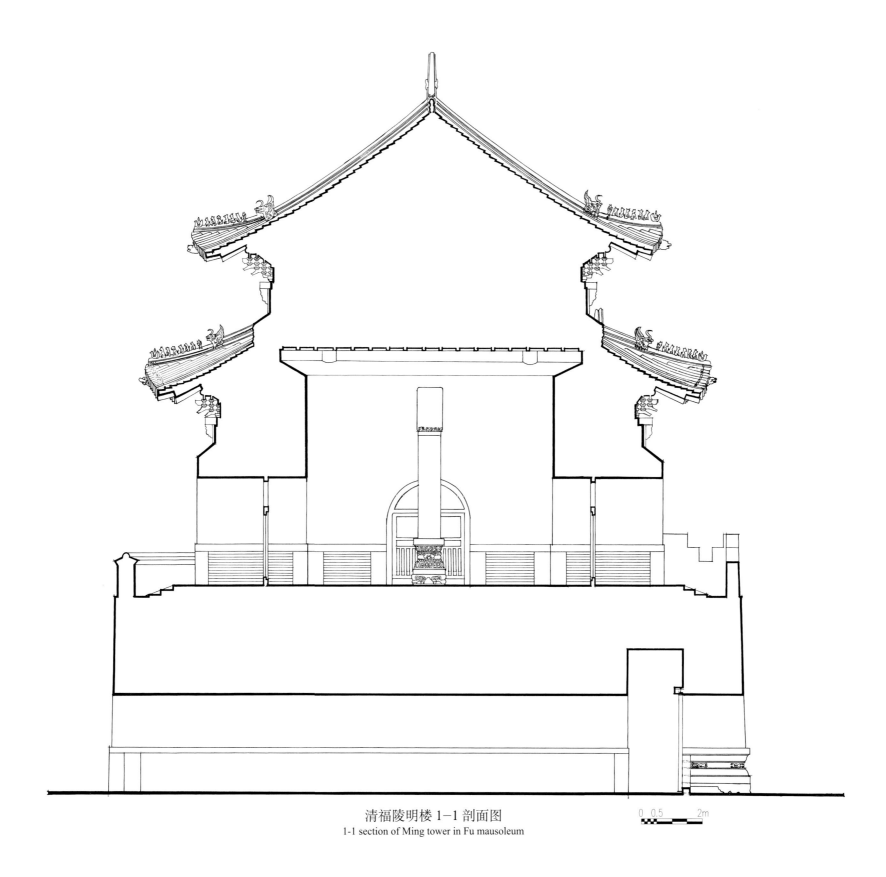

清福陵明楼 1–1 剖面图
1-1 section of Ming tower in Fu mausoleum

0 0.5   2m

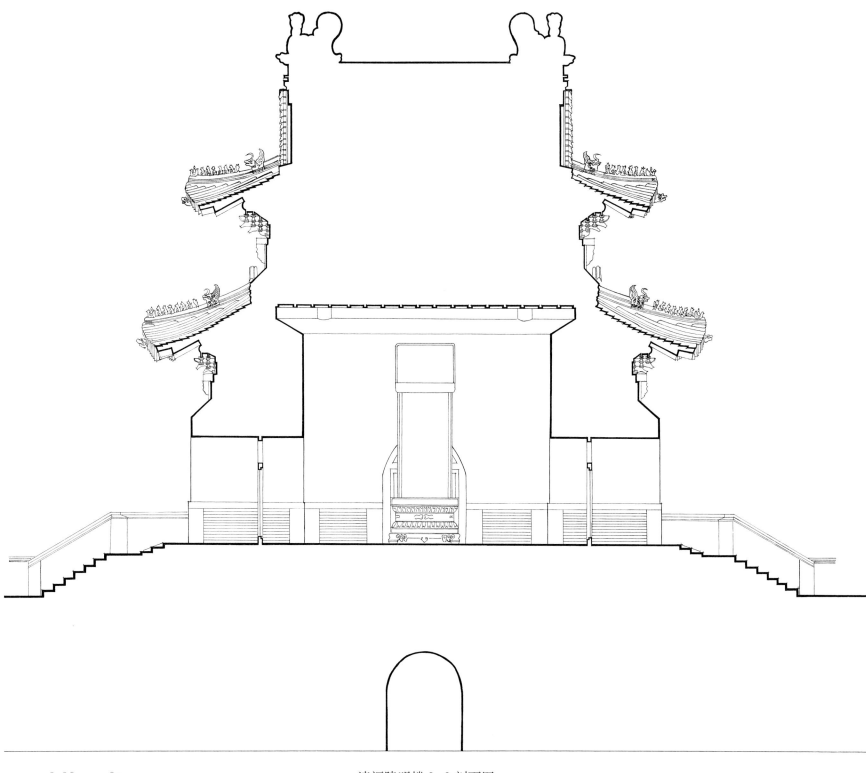

清福陵明楼 2—2 剖面图

2-2 section of Ming tower in Fu mausoleum

清昭陵

Zhao Mausoleum

Project Information

Location: North suburb, Shenyang, Liaoning province

Construction Date: 1643, the eight year of Chongde reign period of the Qing dynasty

Area: 160,000 square meters

Administrative Office: Shenyang City Administration and Enforcement Bureau

Responsible Department: Harbin Institute of Architecture and Engineering (Harbin Institute of Technology)

Survey Time: 1985

项目信息

地　　址　辽宁省沈阳市古城北郊

始建年代　1643 年（清崇德八年）

占地面积　16 万平方米

主管单位　沈阳市城市管理行政执法局

测绘单位　哈尔滨建筑工程学院（现哈尔滨工业大学）

测绘时间　1985 年

1. Brief history

Zhao Mausoleum of Qing Dynasty is the tomb for the second emperor Huang Taiji whose posthumous title was Taizong, and his empress Xiaoduanwen whose family name was Borjigin, together with several imperial concubines from Guanju Palace, Linzhi Palace and Xingqing Palace. It is also known as "The North Mausoleum" because of its location in the north of Shenyang. Built in the 8th year of the reign title Chongde(1643), it saw emperor Huang Taiji buried in it in September of the same year, and a year later got the title "Zhaoling Mausoleum". The empress was buried in it in the 7th year of the reign Shunzhi(1650) and the mausoleum was generally completed in the 8th year (1651). During the reign of emperor Kangxi, the extension construction of the mausoleum was undertaken conventionally thanks to the stable politics and the strengthened economy. The underground palace of Zhaoling Mausoleum was rebuilt in the 2nd year of Kangxi(1663), and the funeral urn of emperor Taizong was settled in, while the throne and memorial tablet were set in the Sacrificial Hall. The posthumous title stele was set up in front of the tomb in 1664, and subsequently the Minglou (the radiant tower which marks the entrance to the underground palace) was built in 1665, the Fangcheng (the square bastion) in 1666, and the Shengong-shengde stele and its pavilion in 1688. A greystone gateway was supplemented during the reign of emperor Jiaqing.

2. General Layout

Zhao Mausoleum is the largest and the most magnificent one among the three Mausoleums outside Shanhaiguan Pass in the beginning of Qing dynasty, which locates at the broad and flat area and is surrounded by the boundary stones in red, white and blue. Its cemetery is enclosed by geomantic red wall that presents regular rectangle with a length of 492 meters from north to south and a width of 328 meters from east to west. Outside the geomantic red wall, there was a duty room on each north side out of the eastern and western red gates (Fig.1).

The part from the Dismounting stele to the front red gate outside the cemetery is the prelude section, and along the central axis from south to north there list a pair of ornamental columns, a pair of stone lions, a three-arched stone bridge and a stone gateway. Two pavilions for dressing people and butchering sacrificial animals, which once stood on both sides between the stone gateway and the front red gate had disappeared, only the "Jing" room used as imperial restroom still exists.

一、历史沿革

清昭陵是清太宗皇太极与孝端文皇后博尔济吉特氏的陵墓，陵内除了葬有帝后外，还葬有关雎宫宸妃、麟趾宫贵妃、衍庆宫淑妃等一批后妃。因地处沈阳城之北，俗称『北陵』。始建于崇德八年（1643年），同年九月葬皇太极于陵内，一周年后确定『昭陵』之名。顺治七年（1650年）皇后入葬，顺治八年（1651年）基本完工。康熙年间，随着政治稳定，经济实力增强，按制扩建祖陵的工程开始实施。康熙二年（1663年）改建昭陵地宫，葬太宗等人于宝宫地宫内，在享殿内设宝座神牌。1664年立谥号碑于陵前，1665年建明楼，1666年建方城，1688年建神功圣德碑及碑亭。嘉庆年间增修青石牌楼。

二、总体布局

昭陵是清初『关外三陵』中规模最大、气势最宏伟的一座，它位于一片广阔平坦的地段，四周设有红、白、青三种颜色界桩。其内陵园由风水红墙围合，呈规则的长方形，南北长492米，东西宽328米。在风水红墙外的东红门、西红门外北侧原各有值房一座（图1）。

图3 《中东铁路大画册》载昭陵主体建筑群

图1 《中东铁路大画册》载昭陵远景图

图4 《中东铁路大画册》载昭陵石牌楼

图2 《中东铁路大画册》载昭陵神道及石像生

Fig.1 A distant view of Zhao mausoleum recorded in *The great album of China eastern railway*

Fig.2 The central sacred path and stone statues of Zhao mausoleum recorded in *The great album of China eastern railway*

Fig.3 The main building groups of Zhao mausoleum recorded in *The great album of China eastern railway*

Fig.4 The stone stele archway of Zhao mausoleum recorded in *The great album of China eastern railway*

The front red gate is the main entrance of the geomantic red wall of the cemetery area. On the path from the front red gate to Fangcheng, there list a pair of southern ornamental columns and six pairs of stone figures of lion, Xiezhi, Kylin, horse, camel and elephant (Fig.2), followed by Shengong-shengde stele pavilion and a pair of northern ornamental columns. Four buildings are located on both sides of the path from the pavilion to the Long'en gate (main gate) of Fangcheng, teahouse and kitchen on the east, fruit room and washhouse on the west.

The bastion-like Fangcheng is the main part of the tomb temples, with turrets around, gate towers on both the north and south sides, the Long'en gate in the south, and the Minglou in the north. The Long'en Hall (Sacrificial Hall) is located near Minglou at inner north of Fangcheng, accompanied by side halls and side towers in front of it (Fig.3), and a two-column gate and stone five-offering table behind it, then vaulted doorway at the back. One can enter the Crescent Court through the vaulted doorway under the Minglou, facing a glazed terra-cotta screen in the front, and step paths on both sides for walking up and down to the Fangcheng. Behind the Crescent Court there are Baocheng(retaining wall) and tumulus with the underground palace inside, after which an artificial hill, Longye Hill, is piled up.

3. Buildings

1) The stone gateway

The stone gateway is located at the independent square platform in front of the front red gate, with the type of non-towering with four columns, three bays and three eaves. Stone lion figures are set attached the columns from front and back, with the double-horn beast figures on the outboard of side columns, thus replace the stone buttress of the columns. Substandard frames are attached from the inner sides of columns, whose upper parts are connected with the small architrave through the pier. A full-length Longmen beam is set on the small architrave of central bay penetrating the central bay and its side bays. There are six intermediate sets and two corner sets of Dougong (bracket set) in the central bay, and four intermediate sets of Dougong in the side bay. The roof is gable-and-hip style with gentle slope and remarkable concave curve. The horizontal tie beams of the stone gateway are in rich carved decorations, except the flat small architrave of the central bay. The columns are without any decoration (Fig.4).

陵园外自下马碑至正红门一段为前奏部分，自南向北沿中轴线依次还列有华表一对，石狮一对，三孔石桥一座，石牌楼一座。石牌楼与正红门之间两侧有更衣亭、宰牲亭两组院落，现两亭已不存，仅有『静房』即御用厕所尚在。

正红门是陵园风水红墙的正门，自正红门至方城一段神道，由南向北依次排列着南华表一对、狮、獬豸、麒麟、马、骆驼和象共六对石象生（图 2），神功圣德碑楼一座，北华表一对，在碑楼与方城的隆恩门之间两侧分列四厢，东侧为茶房、膳房，西侧为果房、涤房。

城堡式的方城是陵寝的主体，四周设角楼，南北设城楼，隆恩门居南，明楼居北，隆恩殿位于方城内北部靠近明楼处，殿前两侧有配殿和配楼（图 3），隆恩殿后有二柱门和石五供，再后是券门。

步入明楼下的券门就进入了月牙城，月牙城正面有琉璃影壁，两侧有『蹬道』可上下方城，月牙城之后是宝城、宝顶，宝顶之内为地宫。宝城之后是人工堆起的陵山——『隆业山』。

三、单体建筑

1. 石牌楼

石牌楼位于正红门前独立的方形月台上，四柱三间三楼柱不冲天式。四柱前后两面的石狮，加上边柱外侧的双角怪兽，代替了夹杆石。三个开间内均贴柱子，设梓框，梓框上端通过云墩与小额枋交接。明间小额枋上设通长的龙门枋贯穿明间与次间。明间平身科斗栱六攒，角科两攒，次间平身科四攒。屋顶为歇山顶，坡度平缓，凹曲显著。除明间小额枋呈素面外，石牌楼横枋雕饰丰富，立柱则不做装饰（图 4）。

图6 《亚细亚大观》载昭陵正红门袖壁蟠龙

图5 《中东铁路大画册》载昭陵正红门

图8 《中东铁路大画册》载昭陵隆恩殿

图7 《亚细亚大观》载昭陵隆恩门

Fig.5 The main red gate of Zhao mausoleum recorded in *The great album of China eastern railway*

Fig.6 The dragon statue on sleeve walls beside main red gate of Zhao mausoleum recorded in *The great view of Asia*

Fig.7 The Longen gate of Zhao mausoleum recorded in *The great view of Asia*

Fig.8 The Longen hall of Zhao mausoleum recorded in *The great album of China eastern railway*

2) The front red gate

The front red gate is a three-arched gate with yellow glazed tile and gable-and-hip roof. It stands on a Sumeru stylobate with three stairs attached on both front and back, which the side belt stones along the central stair end with the squat lion stone, and the side stairs end with drum-shaped bearing stone (Fig.5). The front voussoirs of the central arch are caved with the pattern of two dragons playing with a ball, and the side bay with clouds decoration. There is a flat plaque embedded above each arch. Under the eave there are many components such as glazed terra-cotta flying rafter, short rafter, Dougong, architrave, hanging flower columns and so on. The sleeve walls on both sides of the gate are caved with colored glazed twisted dragon (Fig.6).

3) Long'en Gate (main gate)

The Long'en Gate is a city-gate type tower, located on a high rampart-like platform whose width is 16 meters and depth 14.3 meters. An arched doorway is cut in the central of the platform, with a big glazed board set above the arch. The eave of platform is decorated with wide lined glazed floriations, with crenels on its top. Ramps are attached on both right and left sides of the platform. The gate tower is three-storey high, both width and depth are three bays on each storey with surrounding colonnade. Each storey has the same width of the central bay but different width of side bays reducing from bottom to top in proper. The gable-and-hip roof and two storey-eaves are covered with yellow glazed tiles, and three animal figures are sitting on each diagonal ridge of all storeys without the immortal figure (Fig.7).

4) The Long'en Hall (Sacrificial Hall)

The Long'en Hall is the main building of Zhao Mausoleum, with three bays both in width and depth and a surrounding colonnade. A big platform, where the whole building with yellow glazed tiles gable-and-hip roof, is surrounded by stone railings, and three stairs are attached to its south side. Its Shuyao, the horizontal narrow dado in the central part, is very high, and Xiafang, the lower dado lands directly without cloud-cluster decoration. The panel of the balustrade has no frame edge. The central bay and side bays all use one set of intermediate seven-tier-parallel-arms Dougong between columns (Fig.8). Three sides of the building are closed with thick wall, while in the front, between columns of the second row, partition leafs are set in the central bay and sill wall and sill windows are used in side bays. The timber construction frame is in Lin-jiu system with nine purlins as well as front and back colonnades. The plates supporting the bracket sets rest on the

---

2. 正红门

正红门为单檐歇山黄琉璃瓦顶三孔券门。坐落在须弥座台基上，前后各三出陛，正阶垂带栏杆以蹲狮结束，垂手踏跺的以抱鼓石结束（图 5）。明间券脸石雕二龙戏珠，次间雕云纹。每券上方都嵌一块光素门匾。檐下有琉璃的飞椽、檐椽、斗栱、额枋、垂花柱等构件。门两侧做袖壁，中心嵌五彩琉璃蟠龙（图 6）。

3. 隆恩门

隆恩门是城门型的门楼，位于宽约 16 米、深约 14.3 米的墩台上。墩台正中辟完整的券洞门，门券上方镶大幅琉璃门额。墩台檐部有宽线条琉璃花饰，顶部为品字形垛口，墩台内侧左右设马道。门楼高三层，各层均面阔三间、进深三间、带周围廊。各层明间面阔相同，次间面阔则自下而上依次缩减。歇山屋顶与两层腰檐均用黄色琉璃瓦，顶层饯脊与各腰檐角脊均置三个跑兽而无仙人（图 7）。

4. 隆恩殿

隆恩殿是昭陵的主体建筑，面阔三间，进深三间，带周围廊，单檐歇山琉璃瓦顶，建筑整体坐落在大月台上。大月台环砌石栏杆，南向三出陛，束腰很高，下枋直接着地，无圭角层，栏板无素边。建筑三面包砌厚墙，前檐做金里装修，明间用槅扇，次间用槛墙槛窗。大木构架用九檩前后廊，檩桁式梁架，平板枋与额枋呈『T』字形组合，殿身内砌次间用一攒七踩平身科斗栱（图 8）。建筑三面包砌厚墙，前檐做金里装修，明间用槅扇，上明造，施不规范的苏式彩画。明间后部设佛龛式暖阁，阁内置神座，供奉陵主。

图 10 《中东铁路大画册》载昭陵哑巴院及金刚墙

图 9 《亚细亚大观》载昭陵隆恩门及角楼

Fig.9 The Longen gate and turret of Zhao mausoleum recorded in *The great view of Asia*

Fig.10 The crescent courtyard and Jingang wall of Zhao mausoleum recorded in *The great album of china eastern railway*.

architrave and form T-shaped cross section. Inside the hall all structural components are exposed without ceiling but exerted non-standardized Su-style colored paintings. A niche-like warm room is at the back of the central bay, containing an entablement to sacrifice the owner of the mausoleum.

5) The turret

The turrets on four corners of Fangcheng are two-storied pavilions with gable-and-hip roof and cross ridges, as well as single bay both in width and depth, and surrounding colonnade. Both the full width and depth of the ground floor are 8 meters, and the first floor 7 meters. The width of the bay is the same in both the upper and lower layers, three intermediate sets of Dougong are used between columns. On the ground floor brick wall is built along the second row of columns while on the first floor wood is adopted instead. The roof is covered by yellow glazed tiles with a gourd-shaped vase which is 2.5 meters high and 1.3 meters diameter standing in the center. The vertical, horizontal and diagonal ridges above the gables are all seperated (Fig.9).

6) The Crescent Court

The crescent court is actually an arc wall, in 96 meters length and 7 meters height at the south of Baocheng, forming an enclosed crescent-shaped space nicknamed "Dummy Courtyard" with the north wall of Fangcheng and the platform of Minglou. A Jing'ang wall at the center of the crescent court wall, which is 4.7 meters high , 5.7 meters wide and humping about 0.55 meters, presents as a screen wall with glazed terra-cotta surface and formes the opposite scenery to the archway in the platform of Minglou (Fig.10).

参考文献 References

[1] 王伯扬. 中国古建筑大系（第2册）：帝王陵寝建筑 [M]. 北京：中国建筑工业出版社，1993.

[2] 潘谷西. 中国古代建筑史（第四卷）：元明建筑（第二版）[M]. 北京：中国建筑工业出版社，2009.

[3] 陈伯超，刘大平，李之吉. 中国古代建筑史全集（辽黑吉卷）[M]. 北京：中国建筑工业出版社，2016.

[4] 村田治郎. 亚细亚大观 [J]. 亚细亚写真大观，1942.

[5] 侯幼彬. 读建筑 [M]. 北京：中国人民大学出版社，2012.

[6] 辽宁省图书馆. 盛京风物 [M]. 北京：中国人民大学出版社，2006.

[7] 黑龙江省博物馆. 中东铁路大画册 [M]. 哈尔滨：黑龙江人民出版社，2013.

5. 角楼

方城四隅的角楼均为歇山十字脊的两层重楼，单开间，单进深，带周围廊。底层通面阔、通进深均为8米，二层均为7米。明间面阔上下层相同，均用三攒平身科斗栱。底层沿金里砌砖墙，二层做金里装修。屋面铺黄色琉璃瓦，中心耸立高2.5米、直径1.3米的葫芦状宝瓶。四面山花的垂脊与博脊、戗脊均呈『断砌』的状态（图9）。

6. 月牙城

月牙城为宝城的南段一道长96米、高7米的弧形城墙，它与方城的北壁及明楼墩台围合成一个封闭的弓形空间，俗称『哑巴院』。月牙城弧墙正中设金刚墙，呈影壁形态，高4.7米，宽5.7米，凸起约0.55米，用琉璃饰件贴面，构成明楼墩台门洞的对景（图10）。

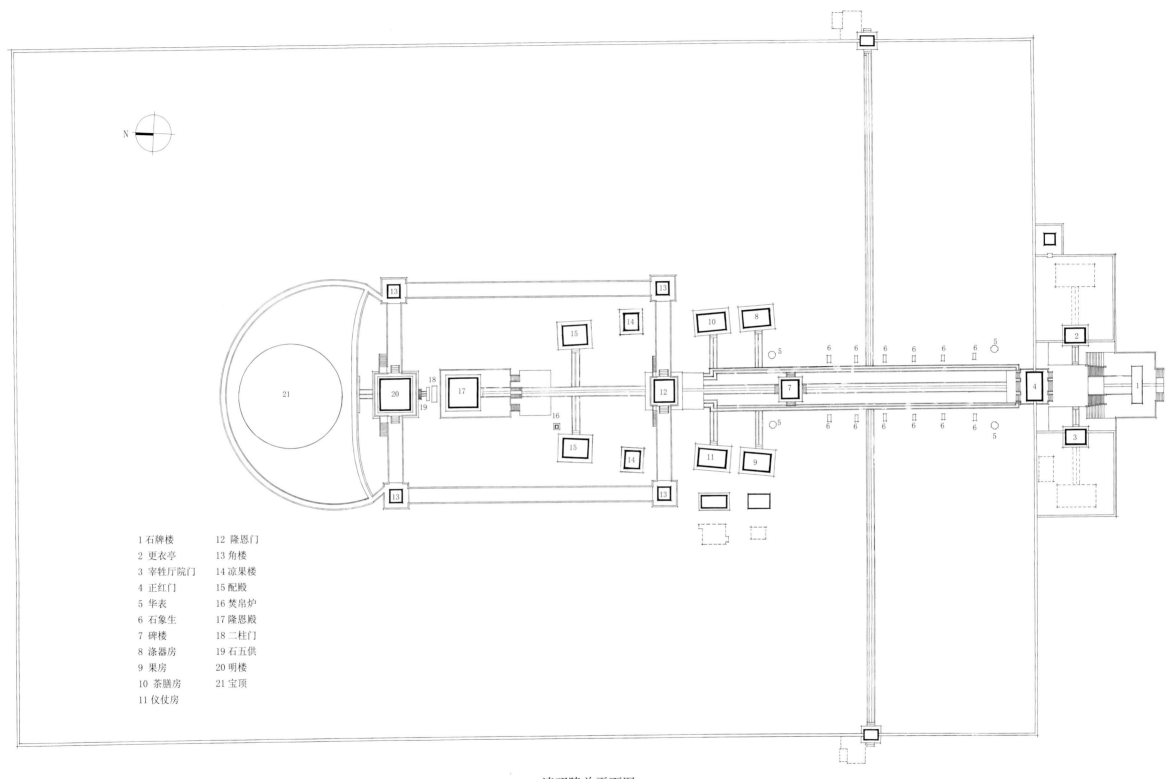

1 石牌楼   12 隆恩门
2 更衣亭   13 角楼
3 宰牲厅院门   14 凉果楼
4 正红门   15 配殿
5 华表   16 焚帛炉
6 石象生   17 隆恩殿
7 碑楼   18 二柱门
8 涤器房   19 石五供
9 果房   20 明楼
10 茶膳房   21 宝顶
11 仪仗房

清昭陵总平面图
Site plan in Zhao Mausoleum

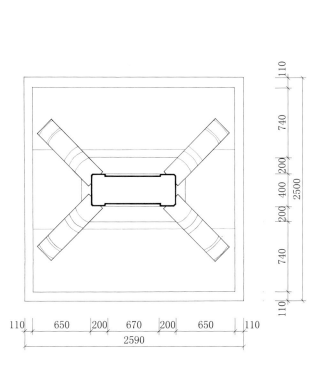

清昭陵下马碑平面图
Plan of Xiama stele in Zhao Mausoleum

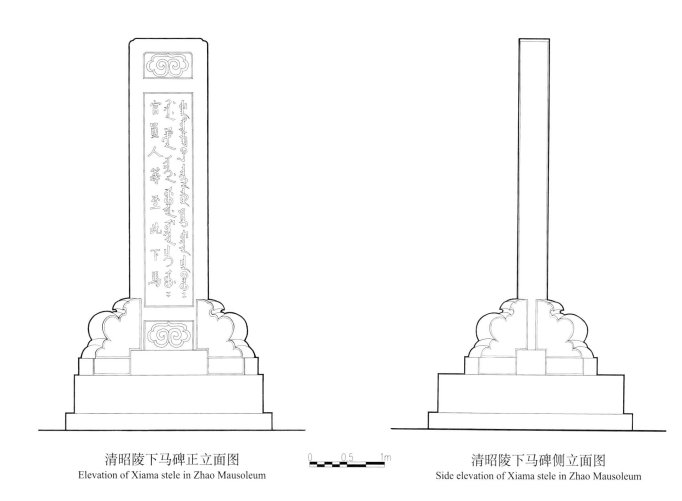

清昭陵下马碑正立面图
Elevation of Xiama stele in Zhao Mausoleum

清昭陵下马碑侧立面图
Side elevation of Xiama stele in Zhao Mausoleum

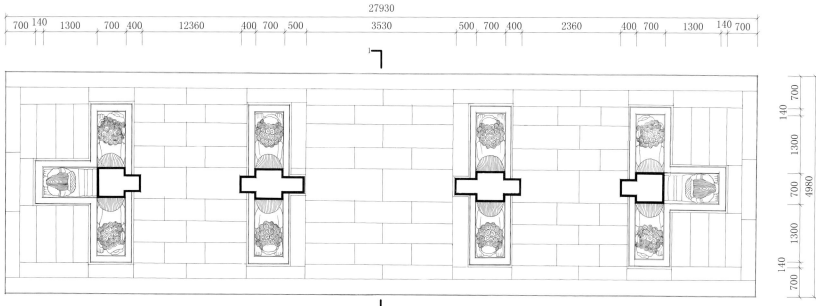

清昭陵石牌楼平面图
Plan of stone archway in Zhao Mausoleum

0   0.5   2m

清昭陵石牌楼正立面图
Elevation of stone archway in Zhao Mausoleum

11.070
10.650
10.260
9.700
8.850
8.500
7.900
7.430
7.300
6.180
5.600
4.880
1.100
±0.000
-0.120

0   0.5   2m

清昭陵石牌楼侧立面图
Side elevation of stone archway in Zhao Mausoleum

清昭陵石牌楼 1-1 剖面图
1-1 section of stone archway in Zhao Mausoleum

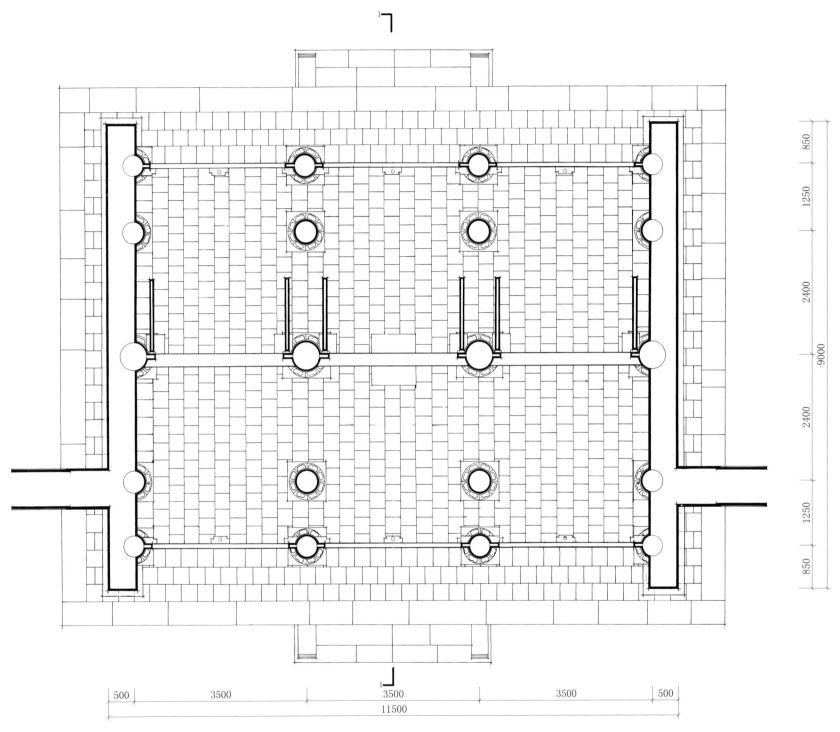

清昭陵宰牲厅院门平面图
Plan of gate on Zaisheng hall in Zhao Mausoleum

清昭陵宰牲厅院门正立面图
Elevation of gate on Zaisheng hall in Zhao Mausoleum

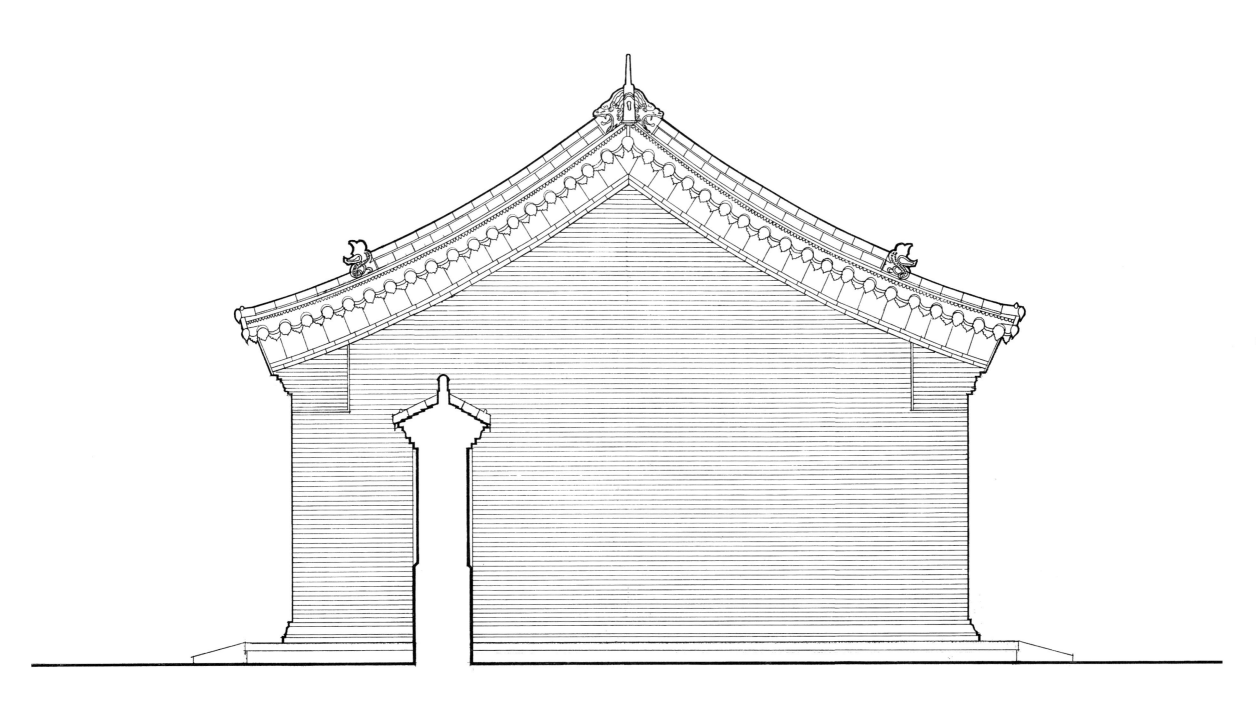

0　　　1　　　　　　3m

清昭陵宰牲厅院门侧立面图
Side elevation of gate on Zaisheng hall in Zhao Mausoleum

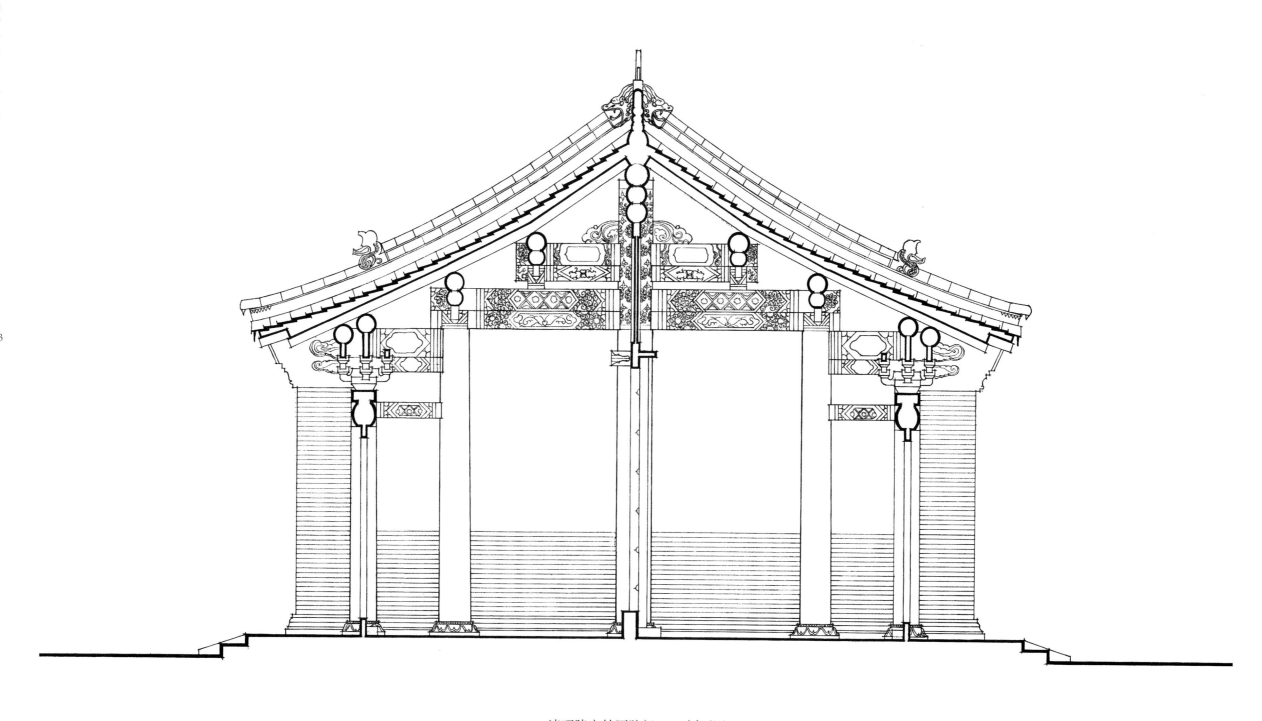

清昭陵宰牲厅院门 1-1 剖面图
1-1 section of gate on Zaisheng hall in Zhao Mausoleum

0 1 3m

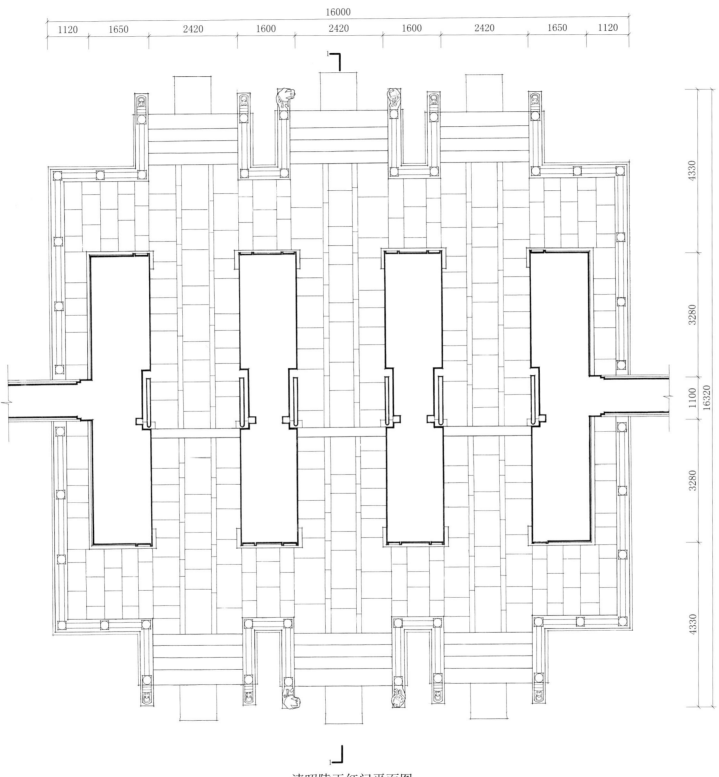

清昭陵正红门平面图
Plan of main red gate in Zhao Mausoleum

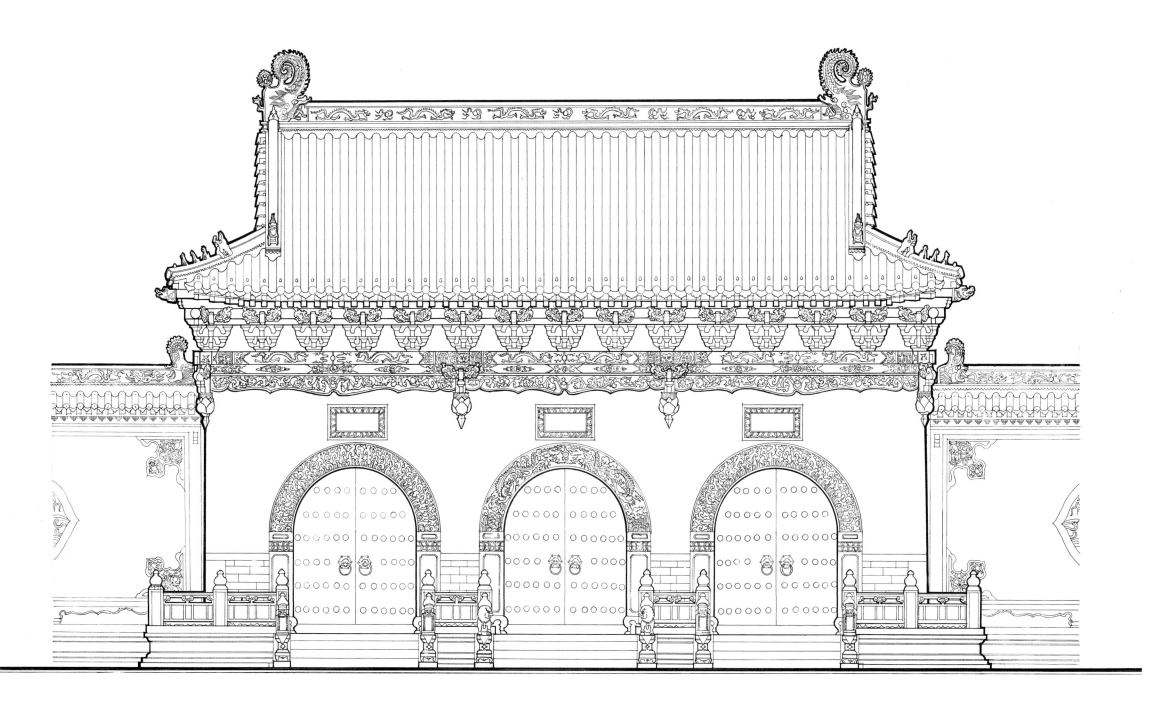

清昭陵正红门正立面图
Elevation of main red gate in Zhao Mausoleum

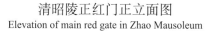

0 1   3m

清昭陵正红门侧立面图
Side elevation of main red gate in Zhao Mausoleum

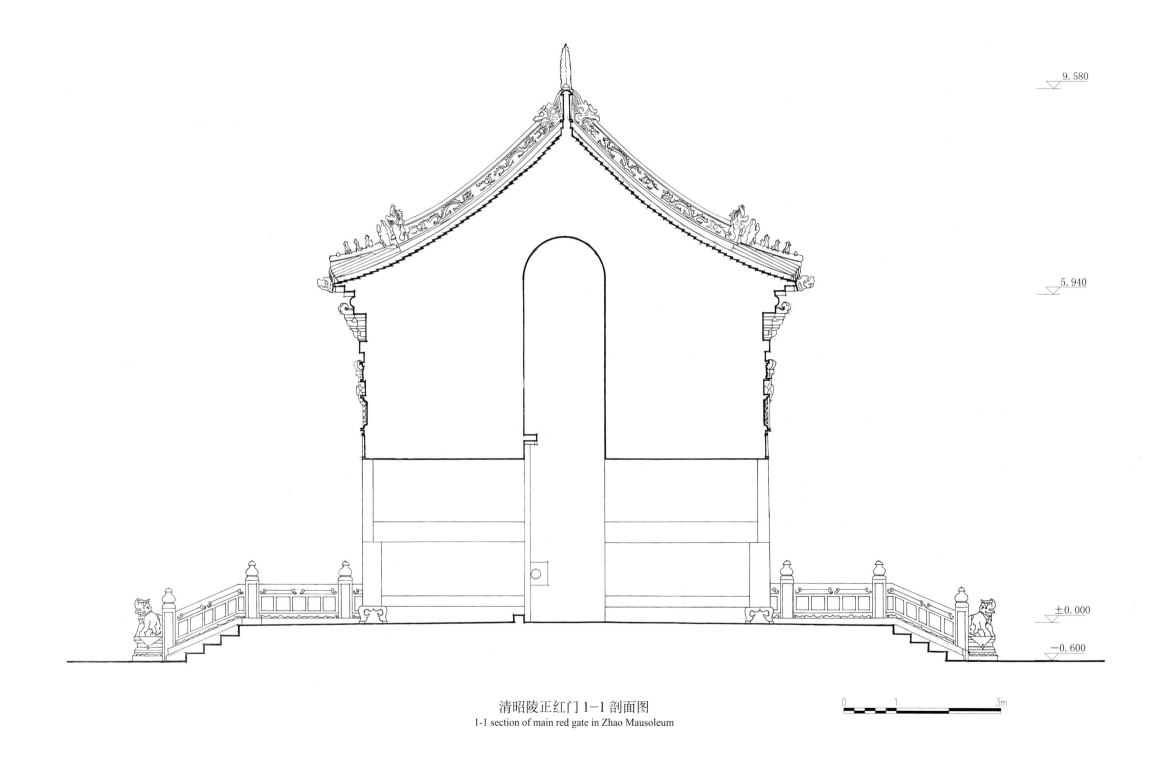

▽ 9.580

▽ 5.940

± 0.000

▽ -0.600

清昭陵正红门 1-1 剖面图

1-1 section of main red gate in Zhao Mausoleum

0   1   3m

0   0.5   1m

清昭陵正红门西袖壁北立面图
North elevation of sleeve wall of main red gate in Zhao Mausoleum

清昭陵隆恩门南华表正立面图
Side elevation of south ornamental column of Longen gate in Zhao Mausoleum

清昭陵隆恩门南华表平面图
Plan of south ornamental column of Longen gate in Zhao Mausoleum

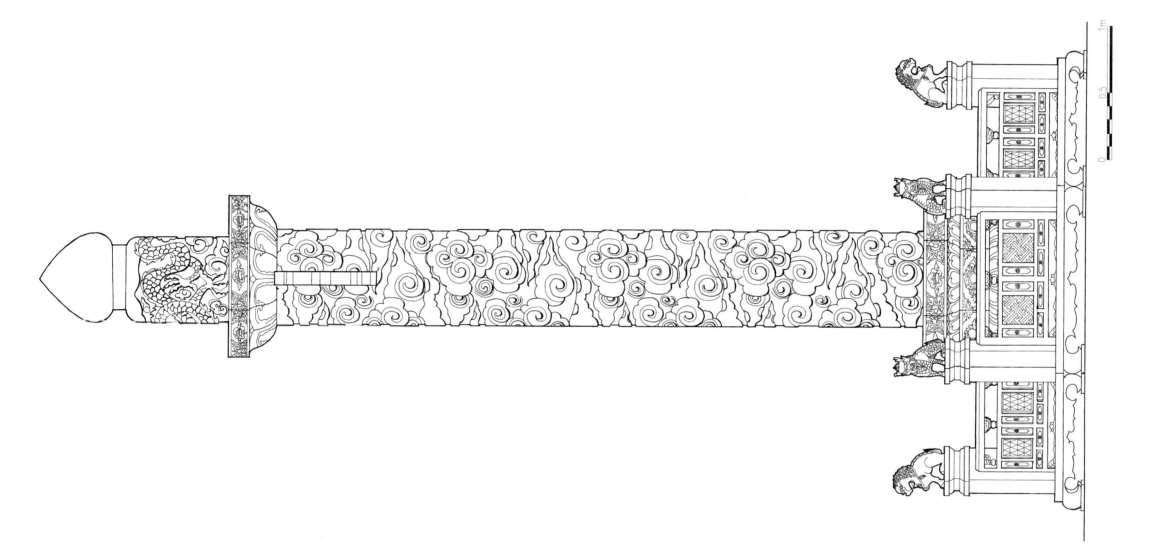

清昭陵隆恩门南华表侧立面图

Side elevation of south ornamental column of Longen gate in Zhao Mausoleum

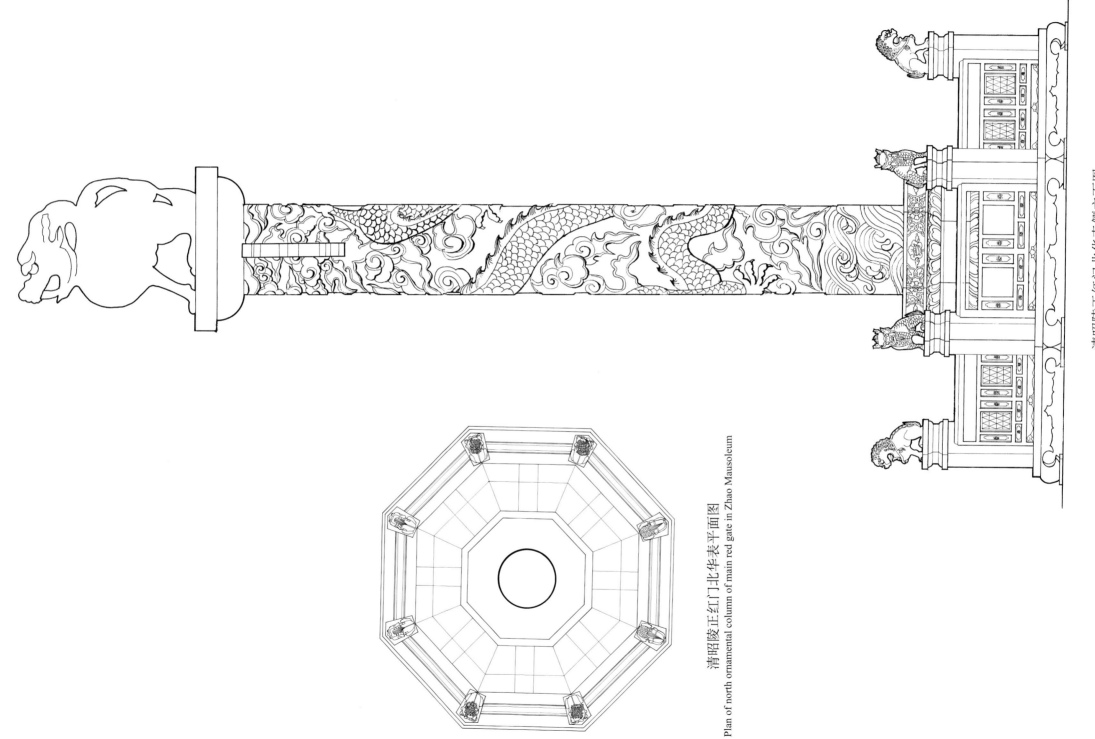

清昭陵正红门北华表侧立面图
Side elevation of north ornamental column of main red gate in Zhao Mausoleum

清昭陵正红门北华表平面图
Plan of north ornamental column of main red gate in Zhao Mausoleum

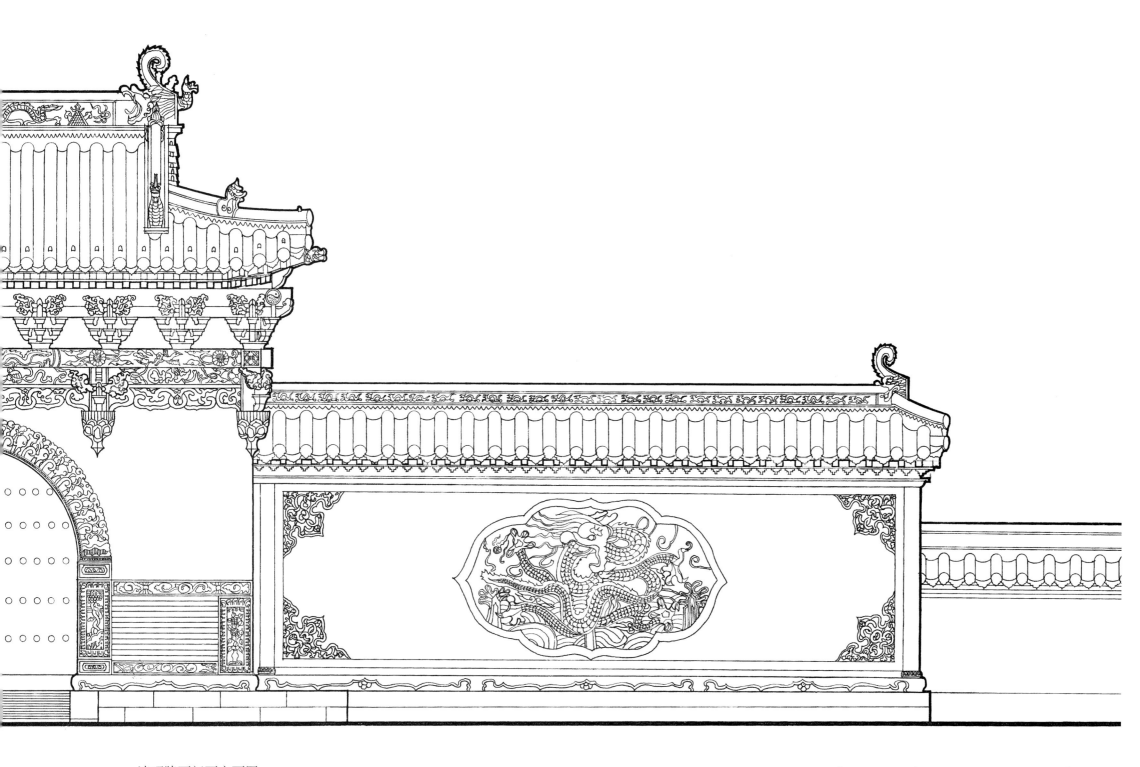

清昭陵西门西立面图
West elevation of west gate in Zhao Mausoleum

0　　　　1　　　　　　　　　3m

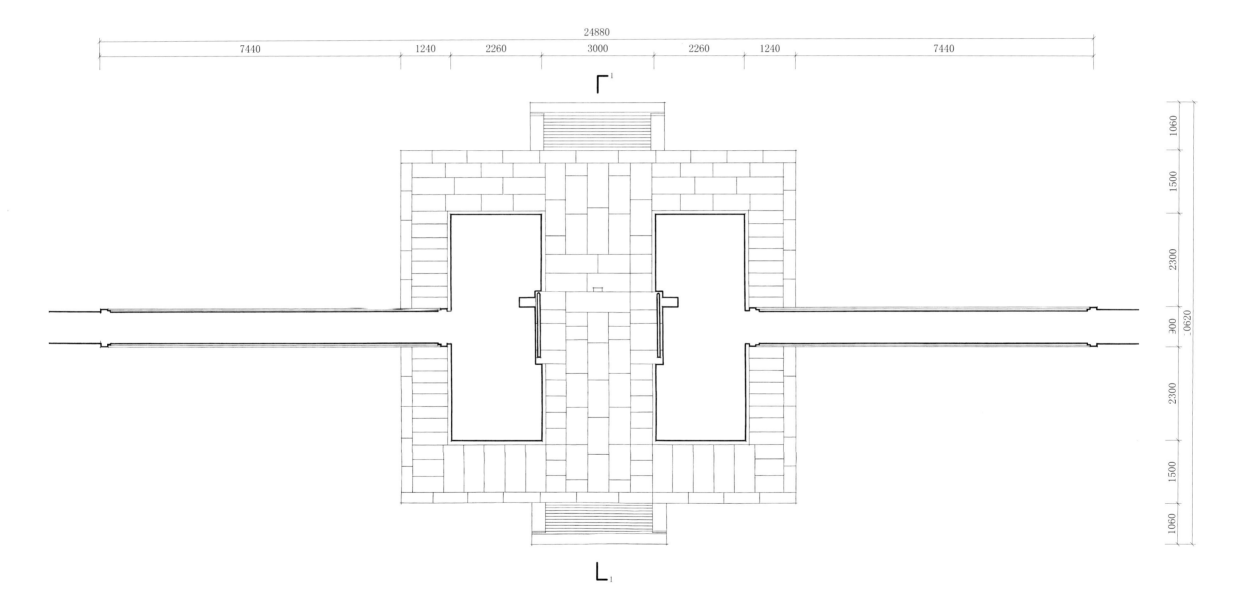

清昭陵西门平面图
Plan of west gate in Zhao Mausoleum

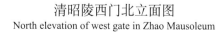

清昭陵西门北立面图
North elevation of west gate in Zhao Mausoleum

0　　　1　　　3m

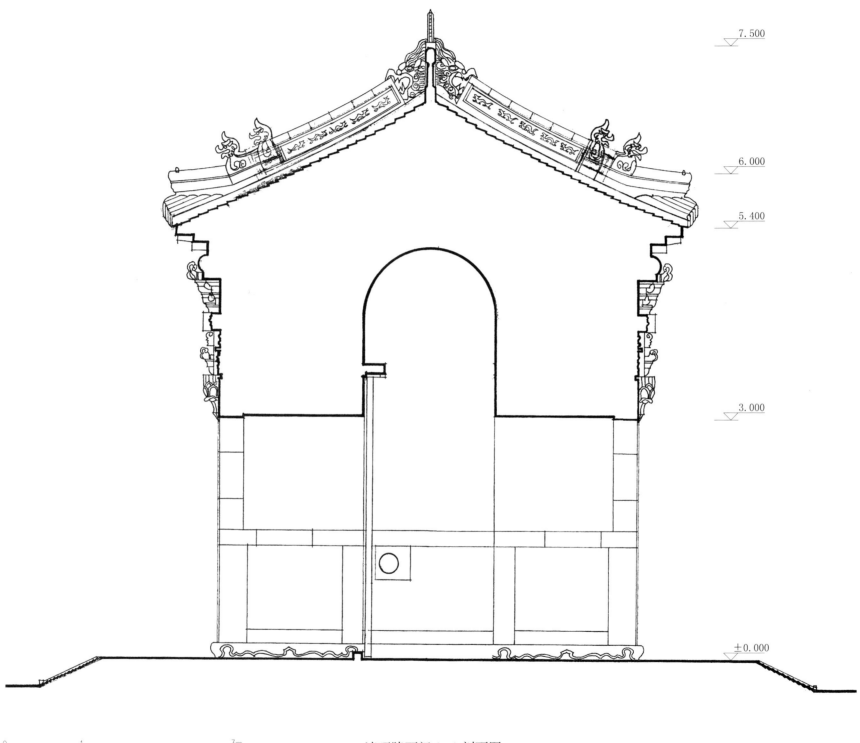

7.500

6.000

5.400

3.000

±0.000

0   1   3m

清昭陵西门 1-1 剖面图
1-1 section of west gate in Zhao Mausoleum

清昭陵西门北袖壁西面龙大样图
Detail drawing of dragon on northwest sleeve wall in Zhao Mausoleum

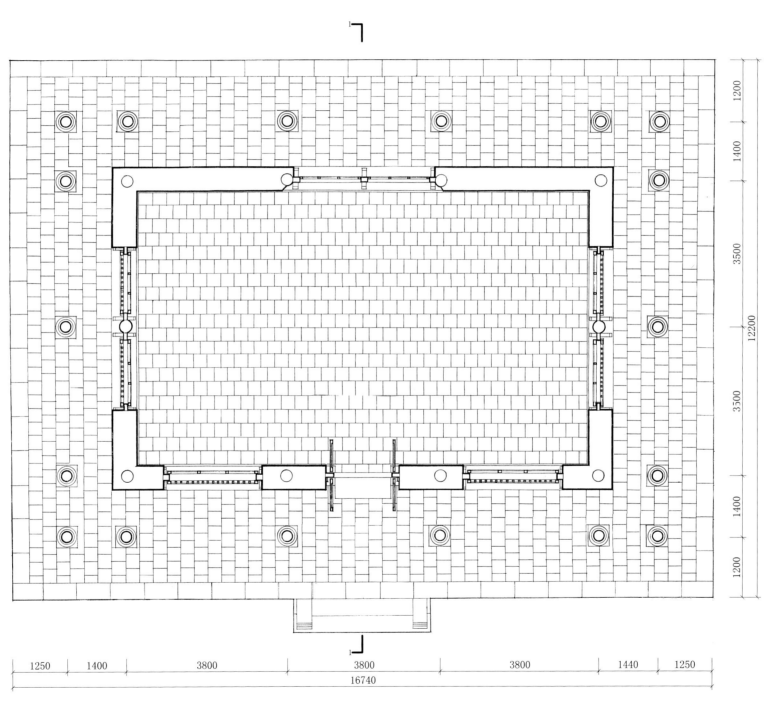

清昭陵茶膳房平面图
Plan of kitchen and washhouse in Zhao mausoleum

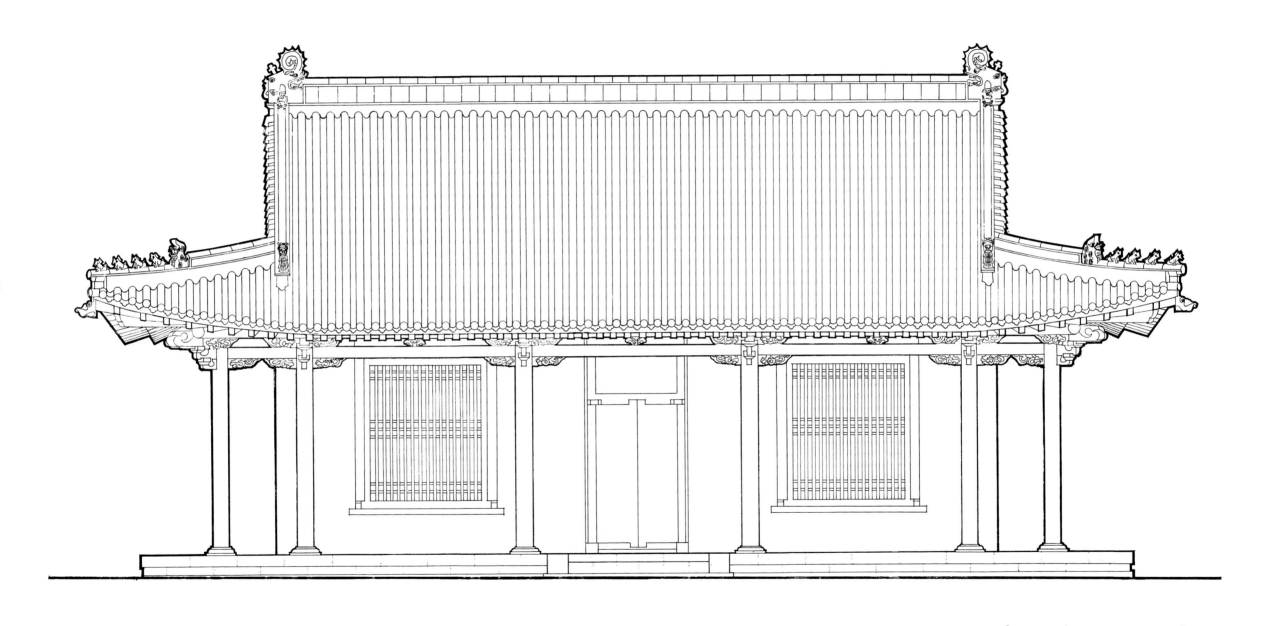

清昭陵茶膳房西立面图
West elevation of kitchen and washhouse in Zhao mausoleum

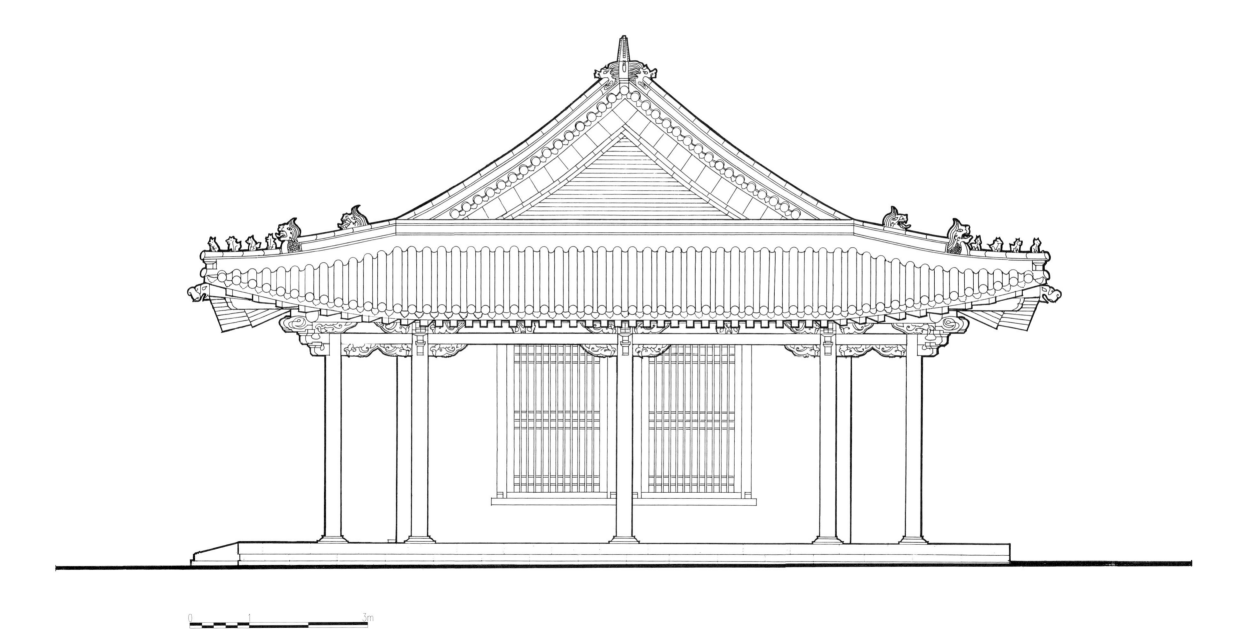

0   1   3m

清昭陵茶膳房南立面图
South elevation of kitchen and washhouse in Zhao mausoleum

8.400

3.340

±0.000
−0.540

−0.100
−0.400

清昭陵茶膳房 1−1 剖面图
1-1 section of kitchen and washhouse in Zhao mausoleum

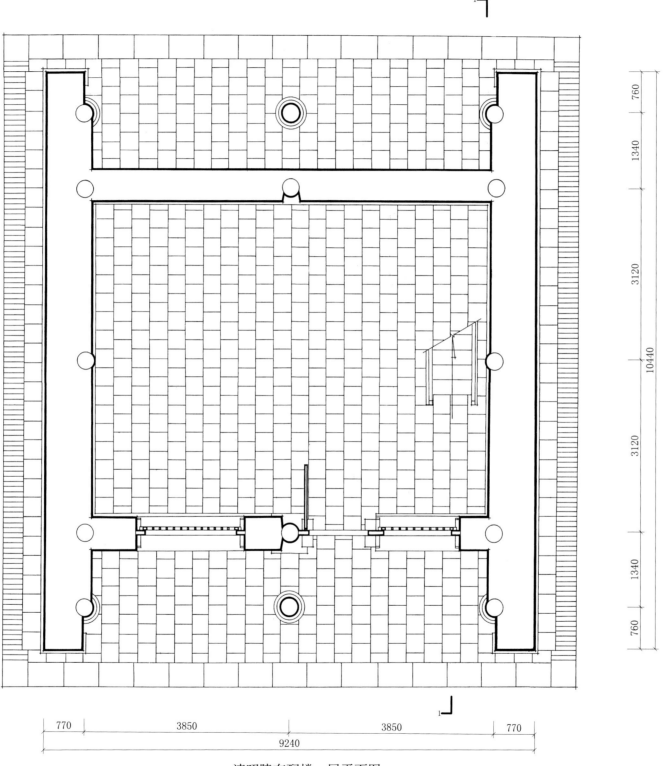

清昭陵东配楼一层平面图
First floor plan of east supporting tower in Zhao mausoleum

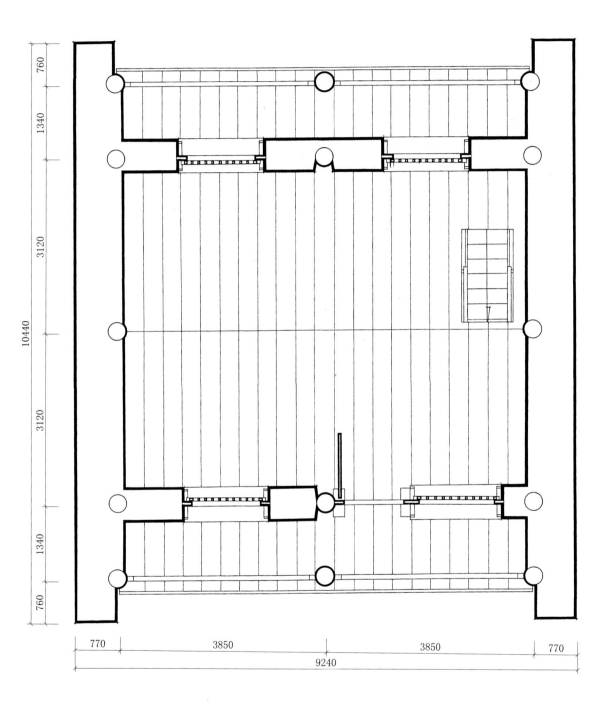

清昭陵东配楼二层平面图
Second floor plan of east supporting tower in Zhao mausoleum

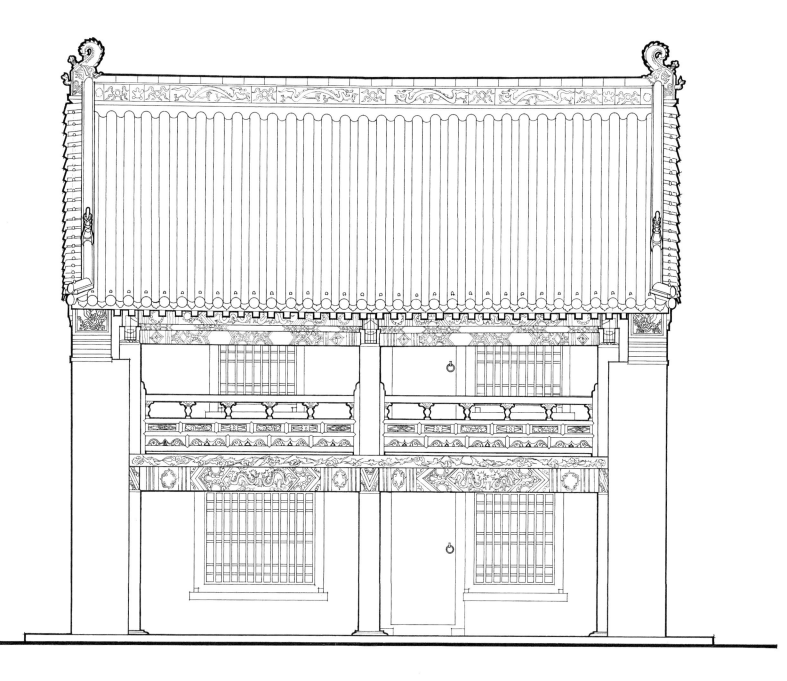

0   1   3m

清昭陵东配楼正立面图
Elevation of east supporting tower in Zhao mausoleum

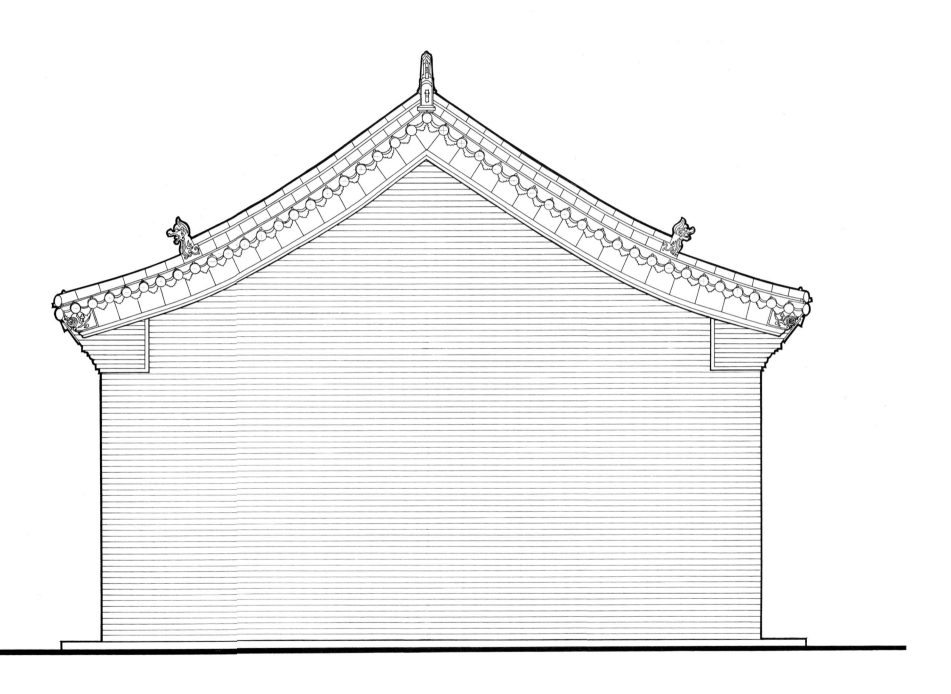

清昭陵东配楼侧立面图
Side elevation of east supporting tower in Zhao mausoleum

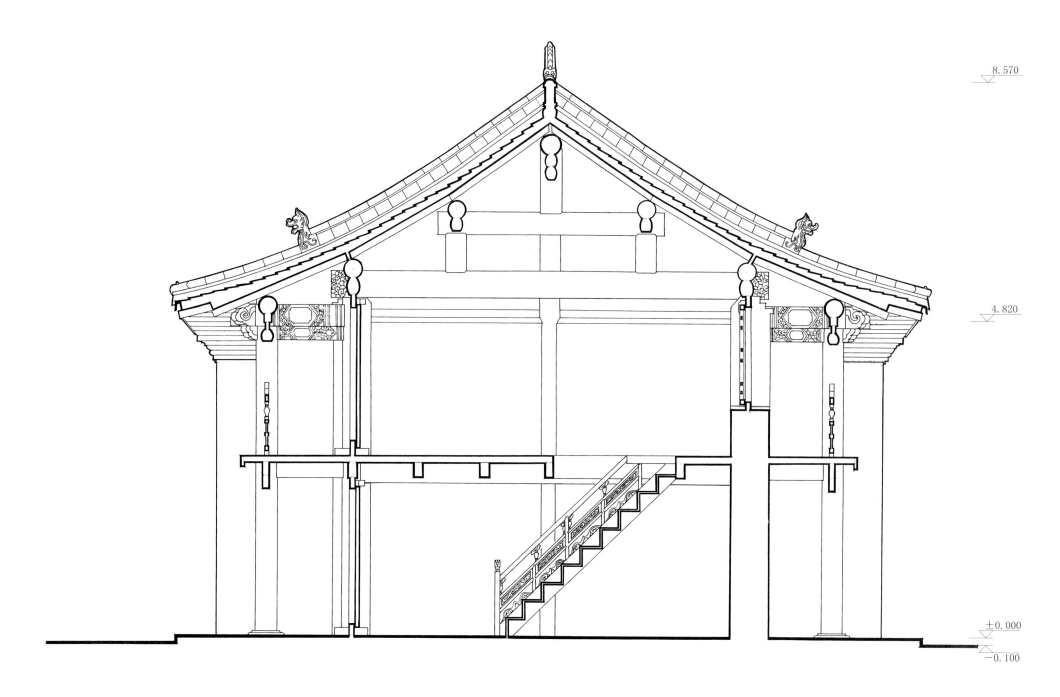

8.570

4.820

±0.000

-0.100

0　　　1　　　　　　3m

清昭陵东配楼 1-1 剖面图
1-1 section of east supporting tower in Zhao mausoleum

清昭陵焚帛炉 1-1 剖面图
1-1 section of stove in Zhao mausoleum

清昭陵焚帛炉平面图
Plan of stove in Zhao mausoleum

500   120 50 140   180   800   180   140 50 120   500
2780

清昭陵焚帛炉正立面图
Elevation of stove in Zhao mausoleum

0　　　　　　0.5　　　　　　1m

清昭陵焚帛炉侧立面图
Side elevation of stove in Zhao mausoleum

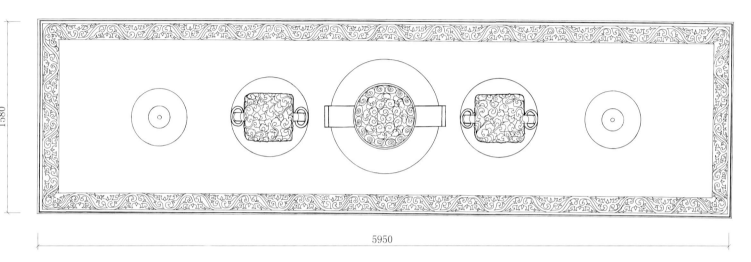

清昭陵石五供平面图
Plan of five stone offerings in Zhao mausoleum

清昭陵石五供侧立面图
Side elevation of five stone offerings in Zhao mausoleum

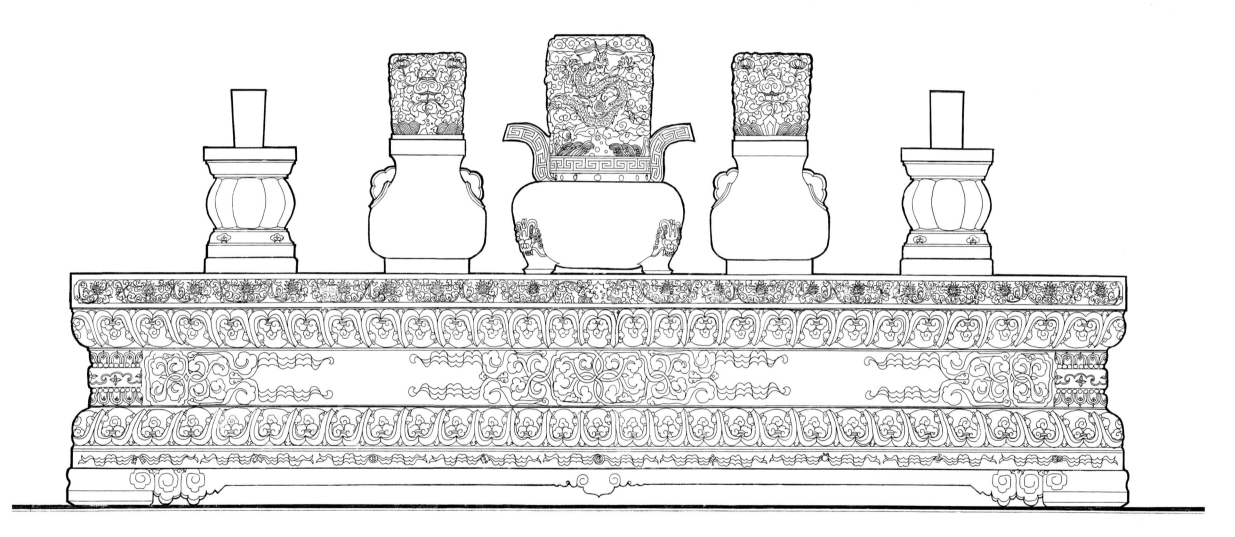

清昭陵石五供正立面图
Elevation of five stone offerings in Zhao mausoleum

0　　　　　0.5　　　　1m

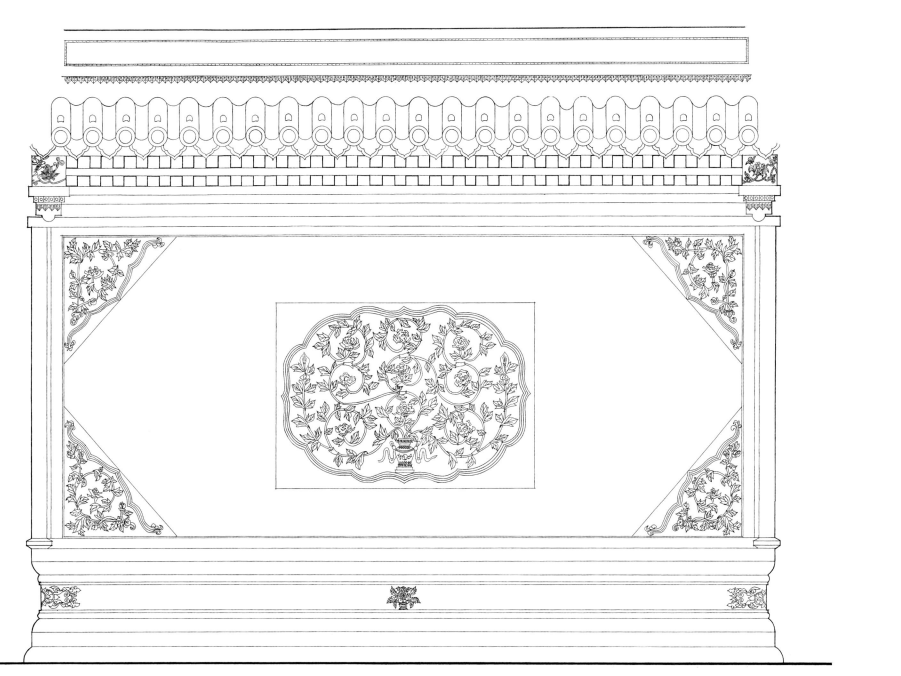

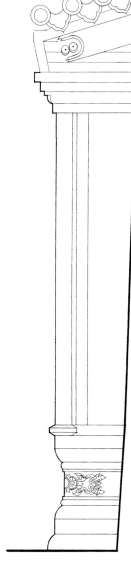

清昭陵金刚墙正立面图
Elevation of Jingang wall in Zhao mausoleum

清昭陵金刚墙侧立面图
Side elevation of Jingang wall in Zhao mausoleum

北
镇
庙

Beizhen Temple

Project Information

Location: West suburb, Beizhen prefecture, Liaoning province

Construction Date: 594, the fourteen year of Kaihuang reign period of the Sui dynasty

Area: 49,800 square meters

Administrative Office: Beizhen prefecture Culture and broadcasting and Television Bureau

Responsible Department: Harbin Institute of Architecture and Engineering (Harbin Institute of Technology)

Survey Time: 1981

项目信息

地　　址　辽宁省北镇市西郊

始建年代　594年（隋开皇十四年）

占地面积　4.98万平方米

主管单位　北镇市文化广电局

测绘单位　哈尔滨建筑工程学院（现哈尔滨工业大学）

测绘时间　1981年

1. Brief History

The Beizhen temple is located on the west hill of Beizhen town, Liaoning Province. It is a mountain temple which was once used by ancient emperor to sacrifice the Yiwulyu mountain, and it's also the only temple which is preserved in good condition among the five different dominating hills of China (Fig.1). The emerging of Beizhen temple has experienced a long history, As recorded in *Zhifangshi of Zhouli*, the northeast part of china has Youzhou, and the dominating hill of Youzhou is Yiwulyu mountain. It also recorded in *Liyi of Suishu* that the first emperor 14 years of Sui dynasty issued an order to build a temple on Yiwulu mountain which named Yiwulyu mountain temple in October of this year.

The present Beizhen temple was initially built in Jin dynasty, and it was repaired and reconstructed several times during the following dynasty. It was extended in large scale in Yongle 19 years of Ming dynasty (A.D.1421), the tablet inscription in *Record of Beizhen Temple Reconstruction* showed that the original form of the temple was removed in this reconstruction, where the new constructions included the front hall with five bays wide, the middle hall with three bays wide, the rear court hall with seven bays wide, the side hall with five bays wide in left and right side of rear hall and Zuoyousi with eleven bays wide in east and west side of the front hall.The enclosing wall, Shenma gate and Zhumen door were built on later so that the present layout pattern was formed basically. The temple was repaired again in Hongwu 23 years of Ming dynasty (A.D.1390), Wawu with three bays wide, Zuoyousi with one bay wide in both side of the original front hall, as well as Zaishengting, Shenku and Shenchu in south direction of Yuan dynasty relics were reconstructed. The reconstruction and repair process were also carried out in Hongzhi 7 years of Ming dynasty (A.D.1494), when the bronze statue recasted, the bell tower and drum tower, the side hall with twelve bays wide in left and right sides, a platform in front of the gate with five room Fangs were newly built.

In Qing dynasty, the repair and renovation of Beizhen temple followed the basic form of Ming dynasty. In Kangxi 46 years (A.D.1707), the temple was restored to the original form. In Yongzheng first year (A.D.1723), it was repaired completely on an imperial order which was token seven years. The record in County Annals of *Beizhen Town* noted that YuXiang hall with seven bays wide, the front hall with seven bays wide, the dressing hall with three bays wide, Neixiang hall with three bays wide and the rest hall with five bays wide were remedied. This restoration is the most completely one in many restoration

一、历史沿革

北镇庙位于辽宁省北镇市城西山冈上，是古代帝王祭祀医巫闾山的山神庙，也是五座镇山中唯一存留完好的镇庙（图一）。北镇庙由来已久，《周礼·职方氏》载：「东北曰幽州，其山镇为医巫闾」；《隋书·礼仪》载：「隋开皇十四年闰十月，诏……北镇医巫闾山就山立祠」，称「医巫闾山神祠」。

今北镇庙始建于金代，后屡经重修与改建。明永乐十九年（1421年）大规模扩建，据《重修北镇庙记》碑载，此次扩建撤其旧制，建前殿五间，中殿三间，后殿七间，在后殿左右各建殿五间，前殿东西各建左右司十一间，增建神马门及外垣、朱门等，基本确立今北镇庙布局。明洪武二十三年（1390年）重修，在元代遗存正殿南建瓦屋三间，左右司各一间，又分别于庙东建宰牲亭、神库、神厨各三间。弘治七年（1494年）维修和扩建，复铸铜像，增建钟、鼓楼及左右翼殿二十间，山门之前扩展平台，台上建坊五楹。

清代北镇庙的修葺遵循明代确立之布局。康熙四十六年（1707年），重修北镇庙，复其原貌。雍正元年（1723年）奉旨大修，七年告成。《北镇县志》载，此番重修后，御香殿七楹、正殿七楹、更衣殿三楹、内香殿三楹、寝宫五楹……清代多次重修中，以此次大修最具规模。乾隆十五年（1750年），

图2 《亚细亚大观》载北镇庙远景

图一 《盛京通志》载北镇庙总平面图

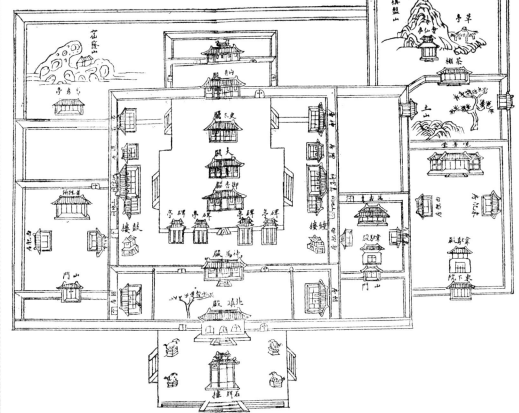

Fig.1 The master plan of Beizhen temple recorded in *Shengjing annals*
Fig.2 A distant view of Beizhen temple recorded in *The great view of Asia*

activities of Qing dynasty. In Qianlong 15 years (A.D.1750), the front hall, rear court hall, god kitchen, god ware house, bell tower and drum tower, stone archway of the temple were all repaired again. In Guangxu 31 year (A.D.1905), the front hall were remedied in bay width from seven to five bays. The front gate in bay width from five to three bays so that the present pattern formed. From late Qing dynasty to early republic china, the sacrificial ceremony of dominating mountain was abolished, so that the Beizhen temple was more and more mutilated. After becoming the third batch of the national cultural relic protection units, new repair work come into operation gradually.

2. General Layout

The Beizhen temple is located on sunny slope of Yiwulyu mountain, where is higher in north part than in south part. The countyard group of the temple is surrounded by red wall, in which the main buildings face south. Its spatial arrangement is gradually spreaded along with the mountain shape. Guangning city can be seen in front of the temple and Yiwulu mountain stands behind the temple (Fig.2).

The main buildings of the temple were arranged on the axis line. Whole spacial sequence along with the terrain goes up, including the stone archway, front gate, Shenma hall, Yuxiang hall, front hall, dressing hall, Neixiang hall and resting hall in order of priority from south to north. A function pattern called "Court is in the front and living quarters are at the rear" are embodied. Axial space of the building group extends along the mountain slope with the trend of step by step penetration. The site has be divided into four platforms along the mountain slope, what are connected with stairs to defuse the height difference among the countyards. On the first platform, a stone archway with five bays wide and six columns is set up, four stone lions in looking very difference are laid around the stone archway. On the second platform, the front gate of Shenma hall can be seen, and the waiting room for officials with five bays wide is located on east and west side of the platform. Shenma hall lays on the third platform, the bell tower and drum tower with gable and hip roof with double eaves are set up behind the Shenma hall symmetrically. There are two tablet pavilions on the way to Yuxiang hall, with some supporting rooms laying on the east and west side wall of platform symmetrically. On the fourth platform, five main halls, such as Xiangyu hall, front hall, dressing hall, Neixiang hall and rest hall, are laid on a large stone foundation which is surrounded by stone rails. These halls constitute the core building group of Beizhen temple (Fig.3). Besides the main buildings and supporting buildings, there are also some subsidiary buildings around Beizhen temple.

二、总体布局

北镇庙高居闾山之阳呈北高南低之势，由红墙围成矩形院落，坐北朝南依山势由低渐高逐层展开空间布局，庙前俯瞰广宁城，庙后遥望医巫闾山（图2）。

寺庙主体建筑群分列中轴线上，随地势上升构成完整空间序列，由南到北串联石牌坊、大红门、神马殿、御香殿、正殿、更衣殿、内香殿、寝殿，呈『前朝后寝』式功能格局。建筑群轴向空间以逐渐加高、层层渗透之趋势在山冈上延展，依山体坡度将场地高差拆解为四层平台，间以踏跺相连，巧妙化解了院落高差，第一层平台设五楼六柱石牌坊一座，石牌坊前后左右分列神态迥异的石狮四座；北上第二层平台，见神马殿正门，平台东西两侧分置朝房五间，向北复行至第三层平台，上置神马殿，殿后左右两侧有重檐歇山钟、鼓楼相对而立，通往御香殿甬路两侧各置碑亭两座，靠平台东西墙两侧各有若干配套用房，左右两侧对称布置；第四层平台置五重大殿，御香殿、正殿、更衣殿、内香殿、

修庙之正殿、后殿、神厨、神库、钟楼、鼓楼、石牌坊。光绪三十一年（1905年）修缮中改原大殿七间为五间，山门五间为三间，形成今北镇庙之规制。清末民初，镇山祭祀仪式废止，庙址渐趋残损，列为第三批全国文物保护单位后修缮工作逐步展开。

图 5 《亚细亚大观》载北镇庙山门

图 3 《亚细亚大观》载北镇庙建筑群

图 4 北镇庙牌坊，作者自摄

Fig.3 The building group of Beizhen temple recorded in *The great view of Asia*
Fig.4 The stone archway of Beizhen temple taken by writer
Fig.5 The front gate of Beizhen temple taken by *The great view of Asia*

Guangning imperial palace is located outside the east wall of the temple, Wanshou temple and Guanyin palace are located outside the east wall of the bell tower, Daxian temple is located outside west wall of the bell tower. These subsidiary buildings have been vanished with only relic plot remained today.

The general layout of Beizhen temple likes nature itself because of the mountains, in which the main buildings distribute along the axial direction, and the supporting buildings are arranged on both side of axial direction symmetrically. The Beizhen temple has both features as a sacrificial architecture in shape and structure, and also as a garden in natural beauty.

3. Individual buildings

The stone archway with five bays wide and six columns is the first building along the axial line of the temple. From the central bay to side bays and final bays, the archway height decreases in proper order. The main body of the archway is constructed of grey sandstone, and several drum stones are placed around the columns (Fig.4). At the front and rear side of the stone archway, there are four beautiful stone lions with the meaning of happiness, angry, sorrow and laughing.

The front gate has a gable and hip roof with single layer, which is covered by green glazed tiles. Chiwen and beast are ornamented on both ends the main ridge and diagonal ridge. There are eave rafter and flying rafter under the roof, but no Dougong. The bottom of the walls is constituted by strip stones. Three arch doors are laid under the front eaves with the word "Beizhen Temple" on a board above the middle door. There are two corner doors on both sidewalls which are surrounded by stylobates and white stone rails (Fig.5).

Shenma hall with five bays wide and three bays deep is used for breeding horses when sacrifice ceremony is hold. The type of its roof is gable and hip roof with single layer, which is covered by grey tiles. There are eave rafter and flying rafter under roof, but no Dougong. The beam frames of the central bay are supported by the middle columns, and the beam frames of the other bays are supported by hypostyle columns. There were two statues of horse and child caring for horse respectively before. The platform in front of the hall is surrounded by stone rails, and two corner doors are laid on each side of the sidewall. Shenma hall used to be as the front gate and was changed to Shanmen hall when a new front gate is built in Hongwu 7 year of Ming dynasty (Fig.6).

寝殿共同矗立在巨大亚字形基座上，环列石栏杆，构成北镇庙核心建筑群（图3）。除主体建筑和配套建筑外，北镇庙还设有附属建筑，寺庙东侧墙外原有乾隆年间所建『广宁行宫』，钟楼东侧原有『万寿寺』及『观音堂』，鼓楼西侧原有『大仙堂』，现基址尚在，建筑无存。

北镇庙总体布局因借山势浑然天成，主体建筑沿南北轴线纵向布置，配套建筑分列中轴线两侧，左右对称，形制相同，布局规则严整，既展现祭祀性礼制建筑的形制特点，又不失自然灵动之美。

三、单体建筑

石牌坊为中轴线上第一座建筑，六柱五间五楼不冲天式，明间、次间、梢间各楼以较大高差依次迭落，牌坊主体由灰色砂岩雕造，各柱前后及边柱外侧置夹柱抱鼓石（图4）。石坊前后置石狮两对，分喜、怒、哀、乐四态，线条精美流畅。

大红门为单檐歇山顶，覆绿琉璃瓦，正脊两端饰吻兽，戗脊上置走兽，檐下设仿木檐椽、飞椽，无斗栱，墙体下碱有压面石和角柱石。正面辟三券拱门，明间券门上置匾额，书『北镇庙』三字。两侧围墙各辟角门，门前及左右置台基，环以白石栏杆（图5）。

神马殿为祭祀时喂养御马之所。单檐歇山顶覆灰瓦，檐下设檐椽、飞椽，无斗栱，面阔五间，进深三间，明间两缝梁架架用中柱，其余各缝用金柱。殿内原有神马及神马童各两尊，现已无存。殿前台

图 7　《亚细亚大观》载北镇庙御香殿

图 6　北镇庙神马殿，作者自摄

Fig.6 The Shenma hall of Beizhen temple taken by writer
Fig.7 The Yuxiang hall of Beizhen temple taken by *The great view of Asia*

Yuxiang hall with five bays wide and three bays deep is used to store the joss sticks and imperial edicts. The type of its roof is gable and hip roof with single layer, which is covered by grey tiles. Dougong printed in three colors are laid under the roof. The central bay of the front and rear eaves and side bays of front eaves are divided by partition board, and the other bays are surrounded by walls. The Wood frame work adopts subtraction column ways, in which hypostyle columns of the central bay and front hypostyle columns of the side bays are removed. Height difference between the front platform and Taiming of this hall is solved by steps, with single stairs between ground and platform, as well as between platform and Taiming in front of the hall.There are Chaoshou stairs on both sides of Taiming (Fig.7). A Fenjin stove with gable and hip roof made of stone is laid on the east up floor under the front platform, and a sundial is set up at west side. Two steles which used to record the sacrifice and sightseeing poems created by the emperor of Qing Dynasty are laid on both east and west side of the platform.

The front hall with five bays wide and three bays deep is used to worship the mountain god and organize sacrifice activities. It has a gable and hip roof with single layer, which is covered by green glazed tiles. The Dougong under the roof is in the style of Shuang'ang Wucai. The front hypostyle columns are subtracted from the beam frames of the central bay, and the rear hypostyle columns are moved half bay backward to the rear eaves. The central bay and side bays under the front eaves are divided by partition boards, and the final bays is divided by sill wall and window sill. The beam frame inside is painted, on which Wenwu hero portraits of Ming dynasty are painted on the walls. There is a Xumizuo made of bricks in the middle of the hall, on which a shrine for Yiwulyu mountain god is set up. The clay sculptures of Wenwu hero are putted on left and right sides under the altar. There are sacrifice steles of Yuan dynasty at east and west side of the front hall (Fig.8).

The dressing hall with five bays wide and two bays deep is used to change clothes when the sacrifice is going on. It has a gable and hip roof with single layer without animal ornament on the ridges, which is covered by grey tiles. There are eave rafter and flying rafter under the roof, but no Dougong. The side bay in width is about one fourth of the central bay. The front and rear eaves of the central bay is divided by partition board, the side bays are divided of sill wall and window sills, and the other bays are enclosed of walls.

Neixiang hall with three bays wide and four rafters deep is used to store the sacrificial offerings and joss sticks. It has a gable and hip roof with single layer, which is covered by grey tiles. There are eave rafter and flying rafter under roof, but no Dougong. The front and rear eaves of the

阶围以石栏杆，殿两侧围墙各开角门。神马殿原为山门，明弘治七年增建山门后，改为神马殿（图6）。

御香殿用以储藏朝廷降香诏书，单檐歇山顶，上覆灰瓦，面阔五间进深三间，檐下斗口单昂三踩斗栱，前檐明间、次间及后檐明间做隔扇，其余各间用墙体围合，大木构架采用减柱造，减去了明间金柱及次间前金柱。殿前月台与台明高差较大，庭院地面至月台及月台至台明均用正阶踏跺单出陛，台明两侧设抄手踏跺（图7）。殿前月台下东侧上层平台设石造歇山顶焚帛炉，西侧石造日晷一座，东西两侧立清代皇帝祭祀、游山诗文碑。

正殿为祭祀医巫闾山神及举行祭典活动的场所，单檐歇山顶覆绿琉璃瓦，檐下斗口双昂五踩斗栱，面阔五间，进深三间，明间两缝梁架减去前金柱，后金柱向后檐移半间，前檐正中三间做隔扇，梢间做槛墙槛窗。殿内梁架彩绘，墙壁绘有明文武功臣像。殿中部有砖砌矩形须弥座，上置神龛，内奉泥塑医巫闾山间山神像，神台下左右各有泥塑文武神像，殿内东西两侧各立有元代御祭碑（图8）。

更衣殿为拜祭山神时更换衣物之所。单檐歇山顶覆灰瓦，无走兽，用檐椽、飞椽，无斗栱。面阔五间，进深两间，梢间面阔约为明间四分之一，前后檐明间做隔扇，前檐次间做槛墙槛窗，其余各间用墙体围合。

内香殿为存放祭品及香火之处。单檐歇山顶覆灰瓦，垂脊上置走兽，设檐椽、飞椽，无斗栱。面阔

图 8　北镇庙大殿，作者自摄

图 9　北镇庙内香殿，作者自摄

图 10　北镇庙寝殿，作者自摄

Fig.8 The great hall of Beizhen temple taken by writer
Fig.9 The Neixiang hall of Beizhen temple taken by writer
Fig.10 The resting hall of Beizhen temple taken by writer

central bay are divided by partition board, the side bays are divided by sill wall and window sill, and the other bays are enclosed by walls (Fig.9).

The rest hall with five bays wide and three bays deep is used to worship mountain god and goddess. It has a gable and hip roof with single layer, which is covered by green glazed tiles. The width from the central bay, side bays to final bay decrease in proper order, on which the number of Pingshenke Dougong among the columns also changes from Sicuan Dougongs, Ercuan Dougongs to Yicuan Dougong. The front eaves of the central bay is divided by partition board. The side and final bays are divided by sill wall and window sills, and the other bays are enclosed by walls. The stair in front of the hall is located on the platform (Fig.10). There is a Xumizuo made of bricks for dedicating to clay mountain god and and god goddess at the central part of the hall.

参考文献
References

[1] 孙大章. 中国古建筑大系（第9册）：礼制建筑 [M]. 北京：中国建筑工业出版社，1993.

[2] 潘谷西. 中国古代建筑史（第四卷）：元明建筑（第二版）[M]. 北京：中国建筑工业出版社，2009.

[3] 陈伯超，刘大平，李之吉. 中国古代建筑史全集（辽黑吉卷）[M]. 北京：中国建筑工业出版社，2016.

[4] 村田治郎. 亚细亚大观 [J]. 亚细亚写真大观，1942，11（124）.

[5] 侯幼彬. 读建筑 [M]. 北京：中国建筑工业出版社，2012.

[6] 清乾隆四十四年奉敕撰. 钦定盛京通志 [M]. 沈阳：辽海出版社，1997.

三间，进深四架椽，前檐明间做隔扇，次间做槛墙槛窗，后檐明间做隔扇，其余各间用墙体围合（图9）。

寝殿是供奉山神及山神娘娘之所。单檐歇山顶覆绿琉璃瓦，面阔五间，进深三间，明、次、梢间面阔依次递减，分置平身科斗栱四攒、两攒、一攒，前檐明间做隔扇，次间、梢间做槛墙槛窗，其余各间用墙体围合，殿前单出陛落于亚字形台基之上（图10）。殿内明间中部置砖砌矩形须弥座，奉泥塑山神及配偶神像。

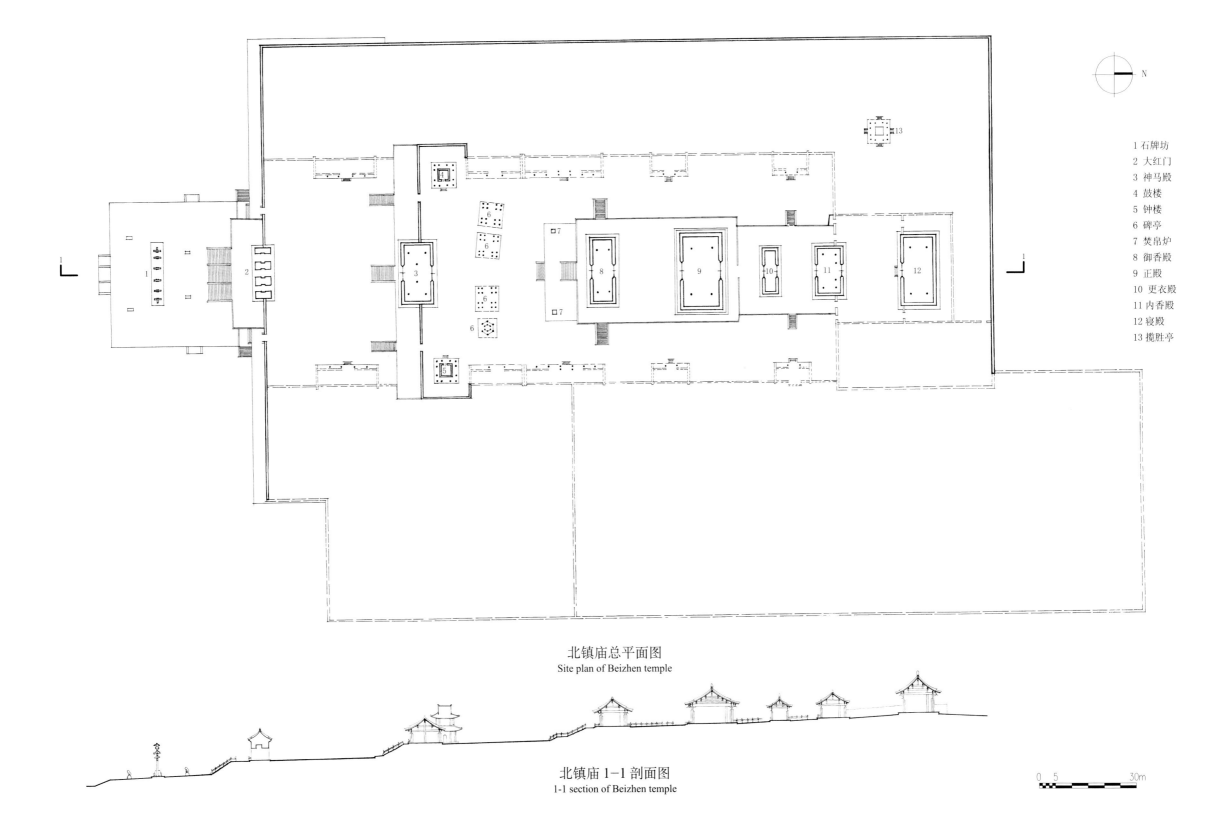

1 石牌坊
2 大红门
3 神马殿
4 鼓楼
5 钟楼
6 碑亭
7 焚帛炉
8 御香殿
9 正殿
10 更衣殿
11 内香殿
12 寝殿
13 揽胜亭

北镇庙总平面图
Site plan of Beizhen temple

北镇庙 1-1 剖面图
1-1 section of Beizhen temple

0 5   30m

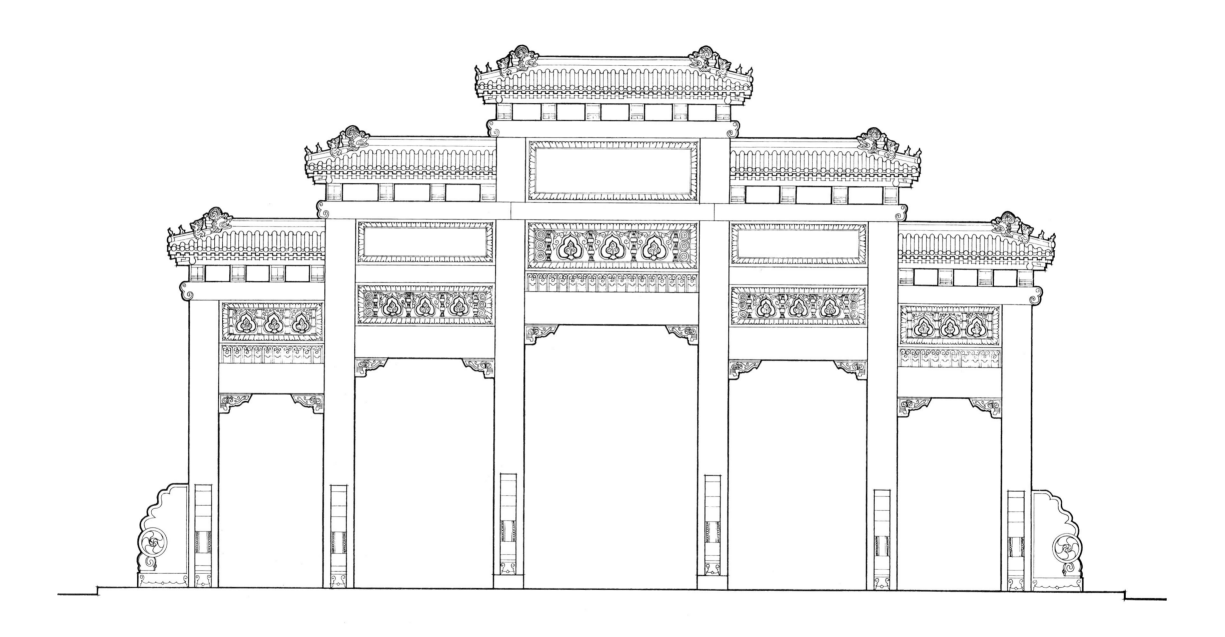

北镇庙石牌坊正立面图
Elevation of stone archway in Beizhen temple

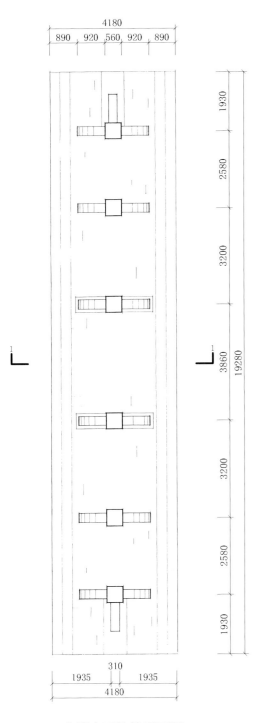

北镇庙石牌坊平面图
Plan of stone archway in Beizhen temple

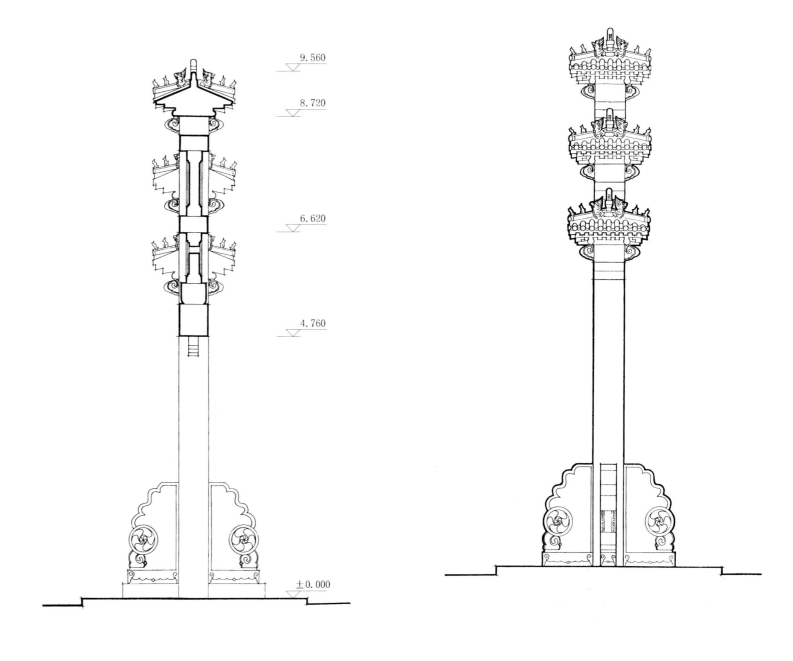

北镇庙石牌坊 1-1 剖面图
1-1 section of stone archway in Beizhen temple

北镇庙石牌坊侧立面图
Side elevation of stone archway in Beizhen temple

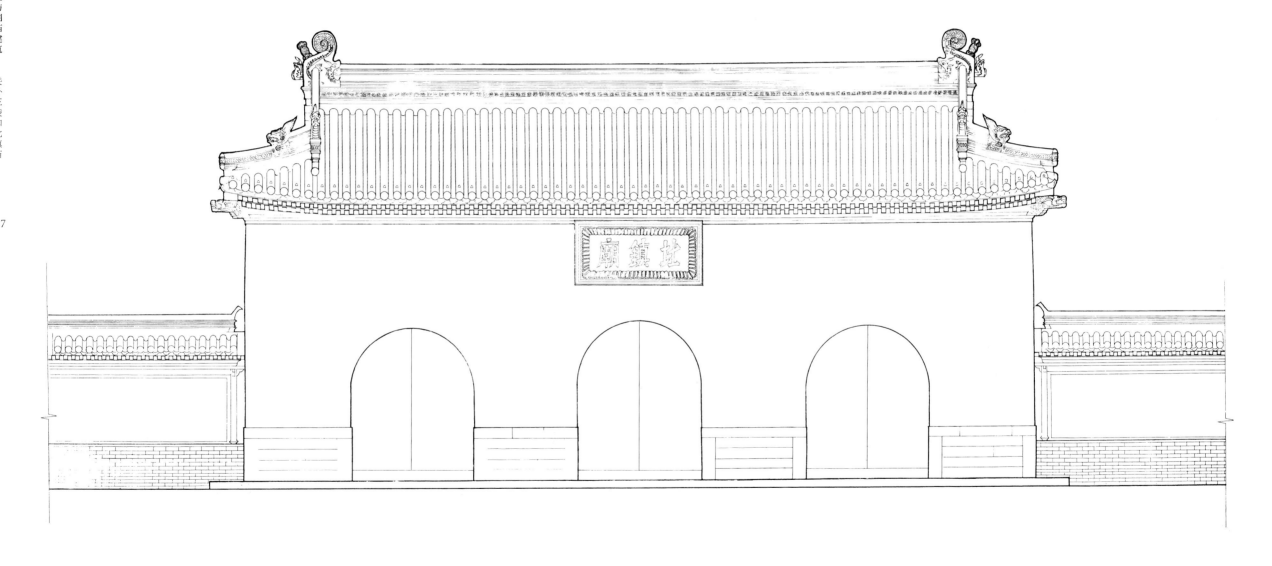

北镇庙大红门正立面图
Elevation of main red gate in Beizhen temple

0 0.5   3m

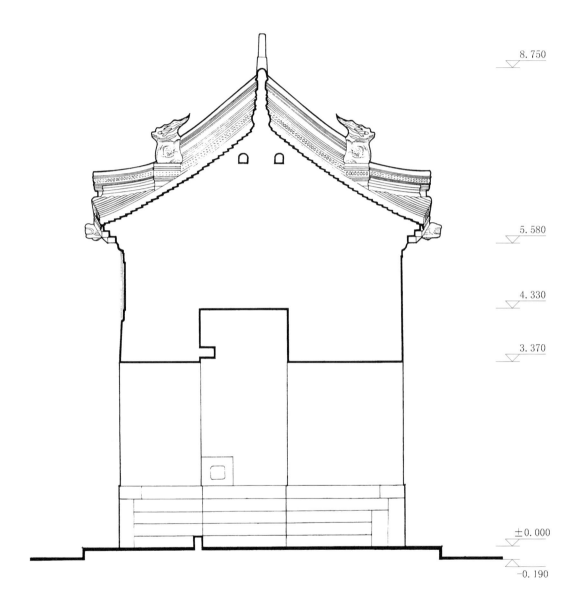

8.750

5.580

4.330

3.370

±0.000

−0.190

北镇庙大红门侧立面图
Side elevation of main red gate in Beizhen temple

0 0.5 3m

北镇庙大红门 1—1 剖面图
1-1 section of main red gate in Beizhen temple

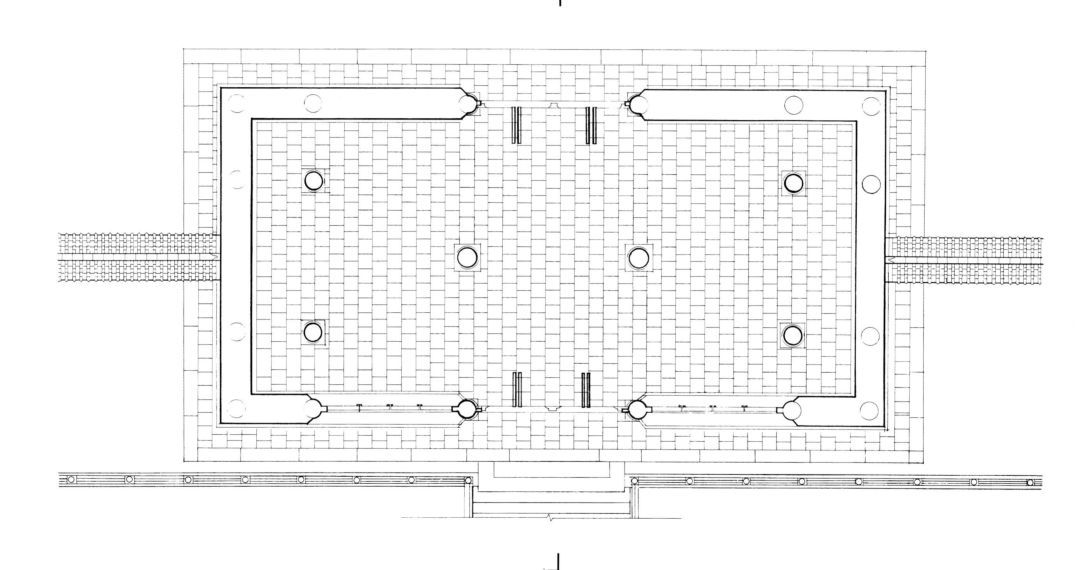

北镇庙神马殿平面图
Plan of Shenma hall in Beizhen temple

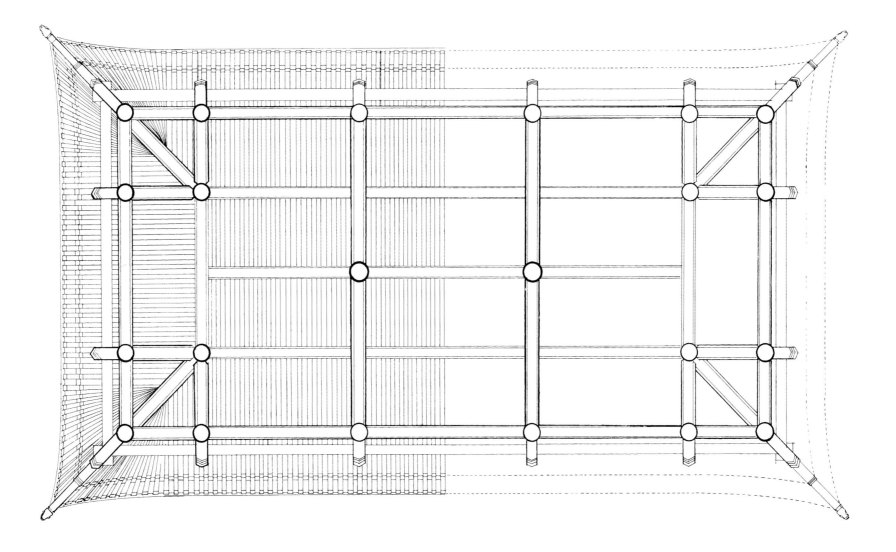

0   0.5   3m

北镇庙神马殿梁架仰视图
Beam frame upward plan of Shenma hall in Beizhen temple

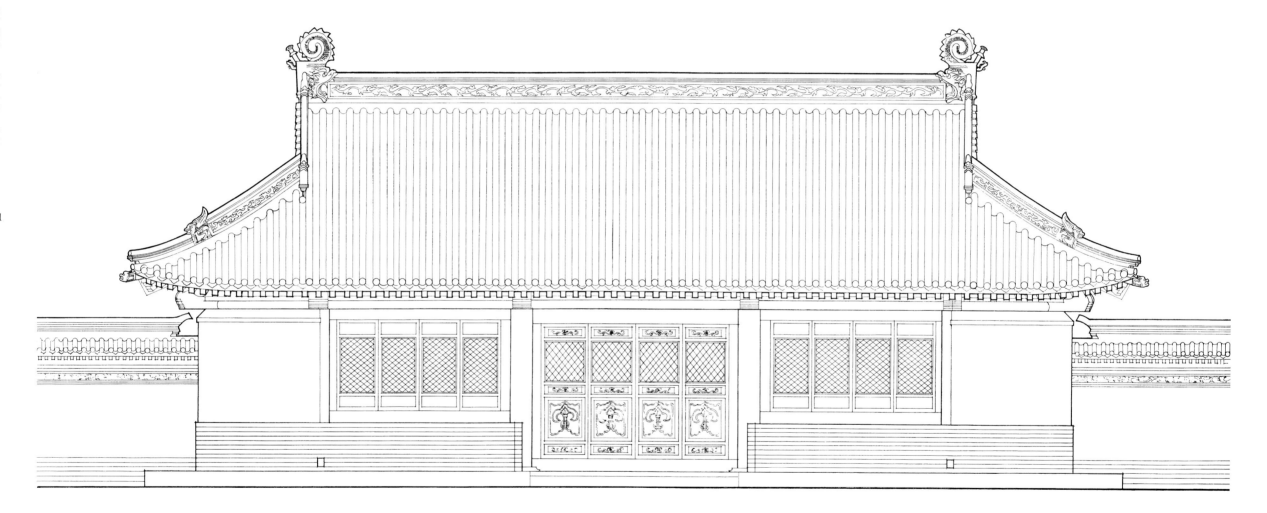

北镇庙神马殿正立面图
Elevation of Shenma hall in Beizhen temple

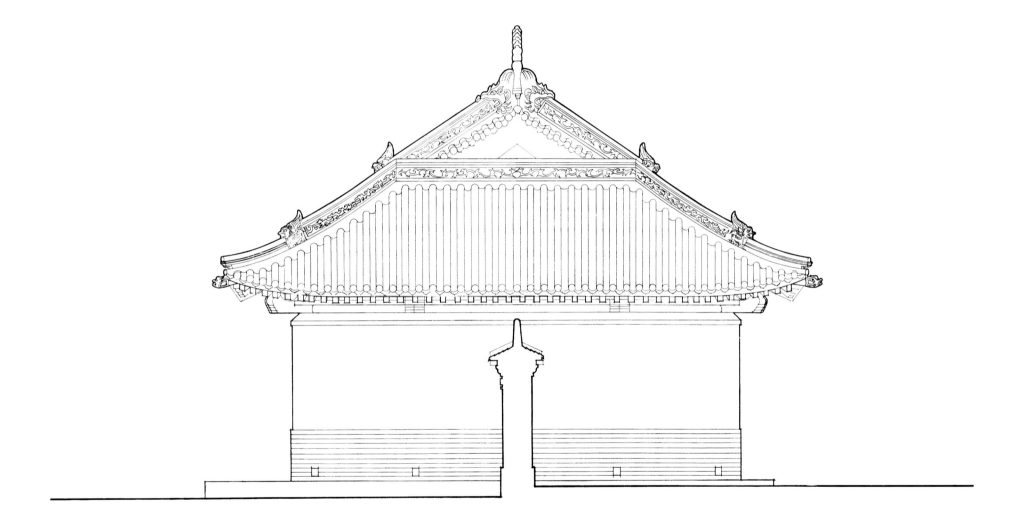

0 0.5   3m

北镇庙神马殿侧立面图
Side elevation of Shenma hall in Beizhen temple

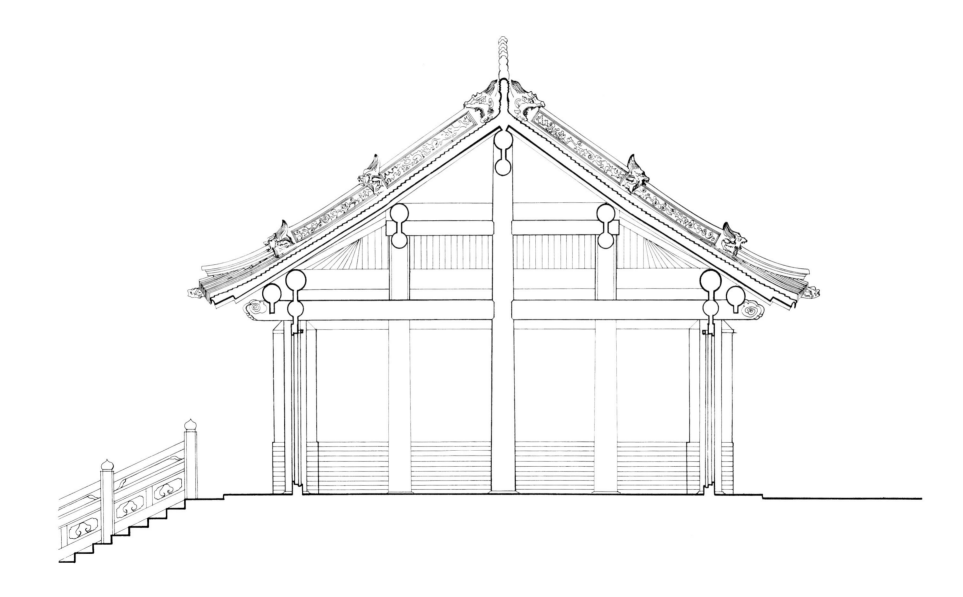

北镇庙神马殿 1-1 剖面图
1-1 section of Shenma hall in Beizhen temple

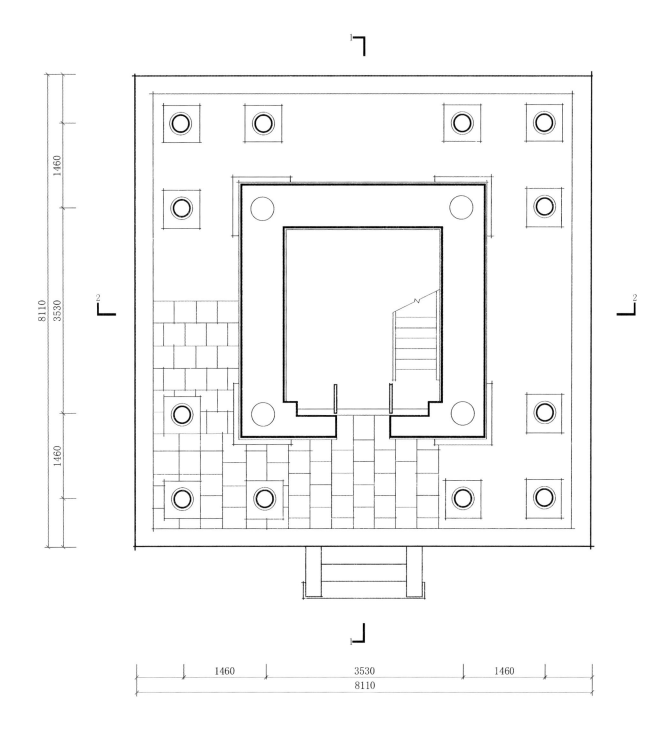

北镇庙钟楼一层平面图
First floor plan of bell tower in Beizhen temple

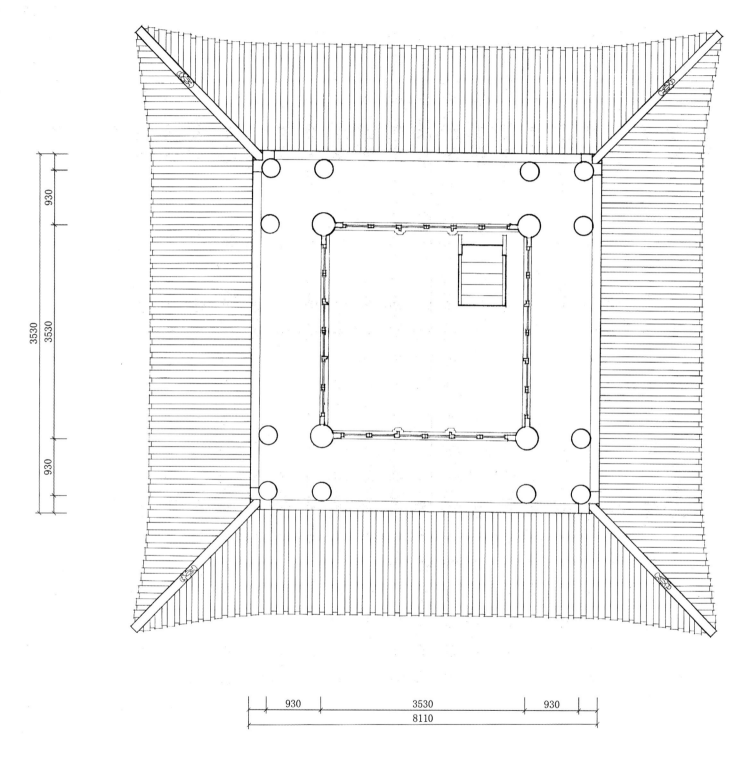

北镇庙钟楼二层平面图
Second floor plan of bell tower in Beizhen temple

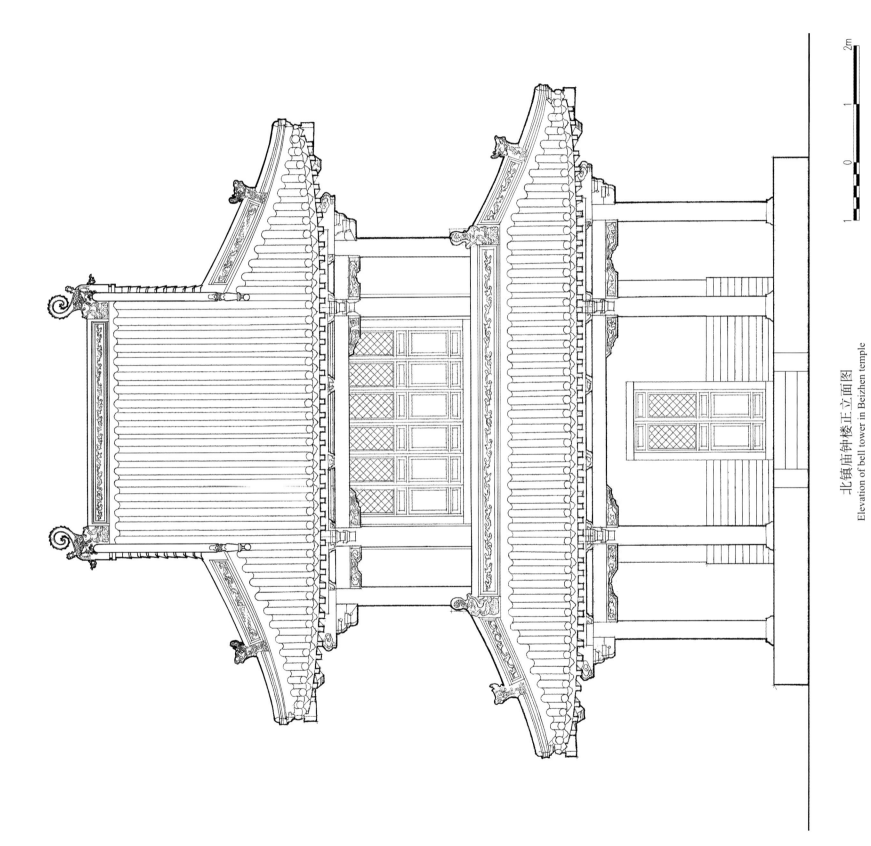

北镇庙钟楼正立面图

Elevation of bell tower in Beizhen temple

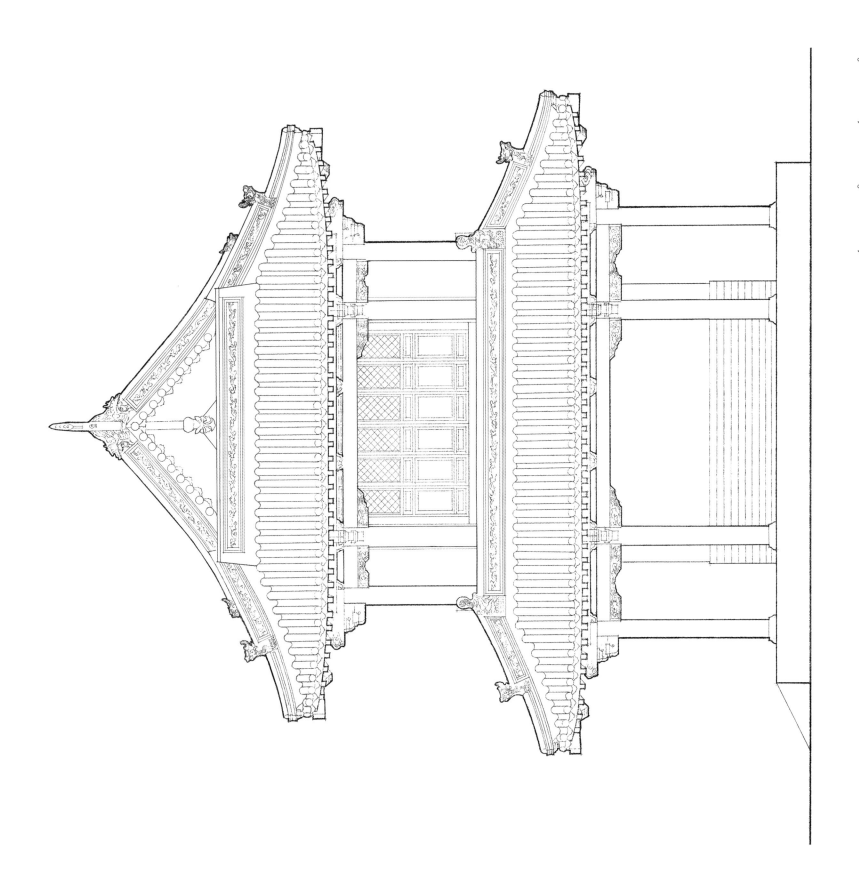

北镇庙钟楼侧立面图
Side elevation of bell tower in Beizhen temple

2m

1

0

1

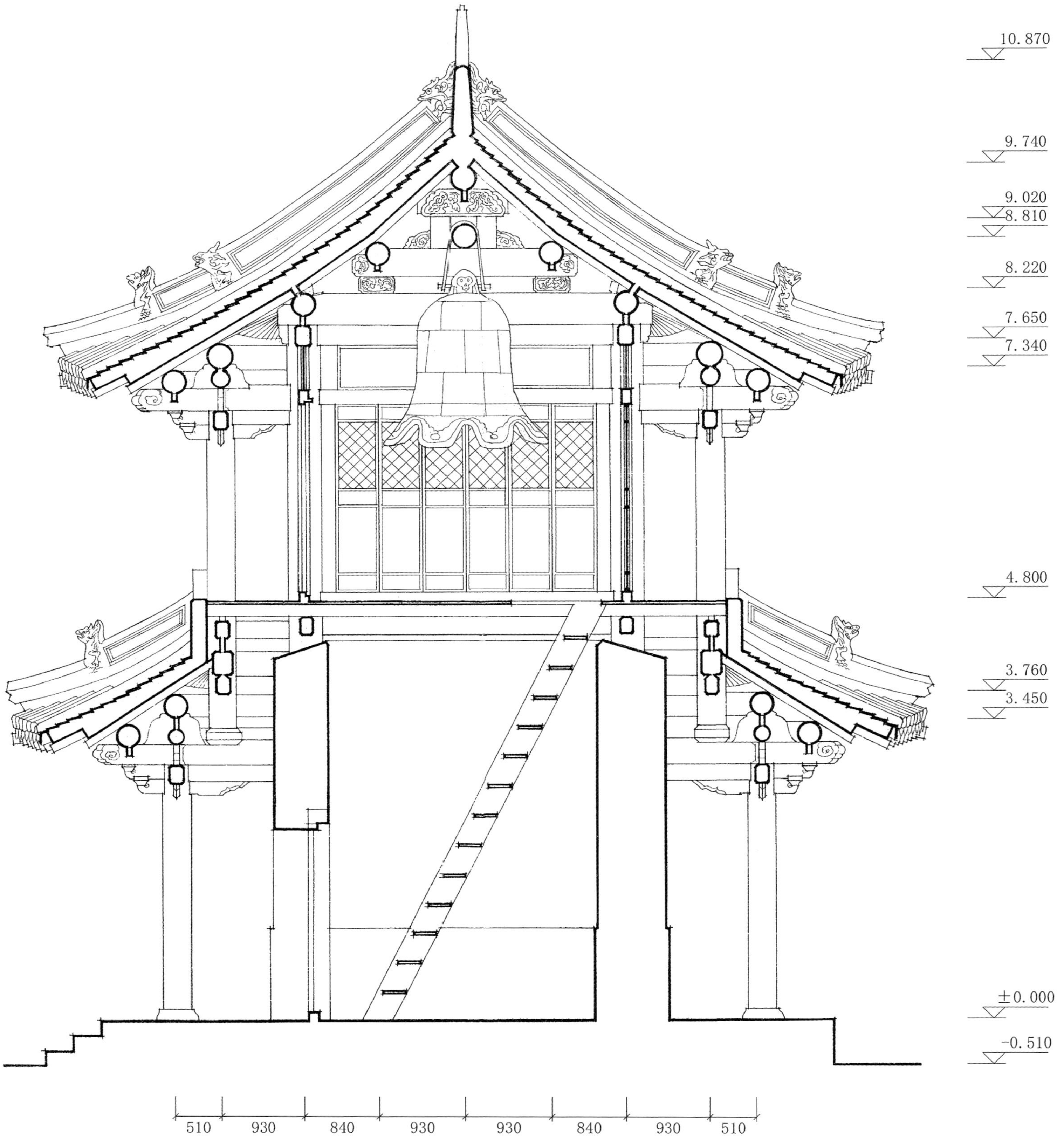

北镇庙钟楼 1—1 剖面图
1-1 section of bell tower in Beizhen temple

10.870
9.740
9.020
8.810
8.220
7.650
7.340
4.800
3.760
3.450
±0.000
-0.510

510　930　840　930　930　840　930　510

2m
0
1

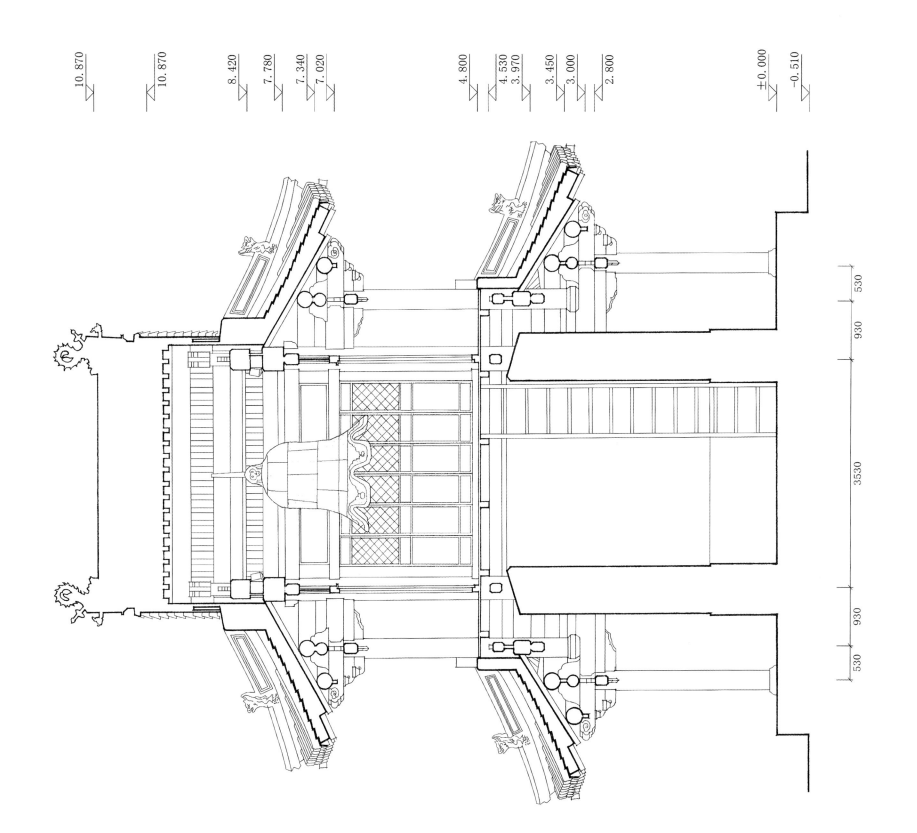

10.870

10.870

8.420

7.780

7.340

7.020

4.800

4.530

3.970

3.450

3.000

2.800

±0.000

-0.510

530

930

3530

930

530

2m

1 0 1

北镇庙钟楼 2-2 剖面图
2-2 section of bell tower in Beizhen temple

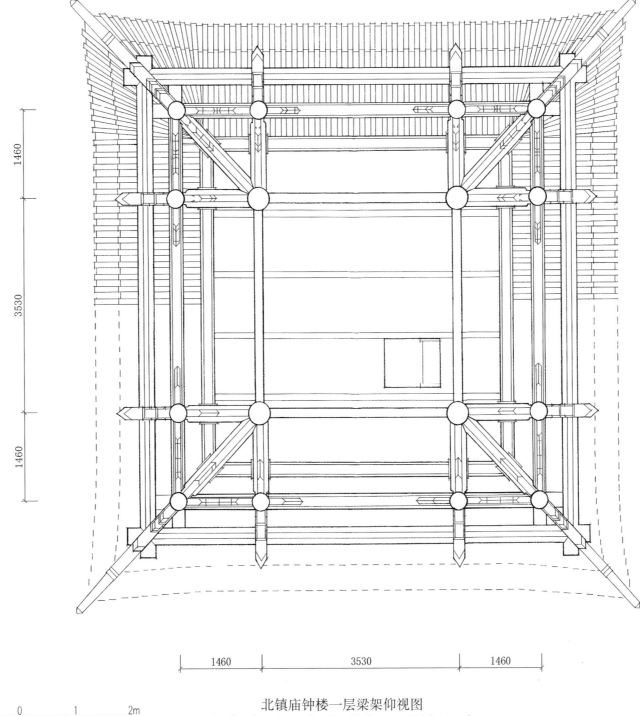

北镇庙钟楼一层梁架仰视图
First floor beam frame elevation of bell tower in Beizhen temple

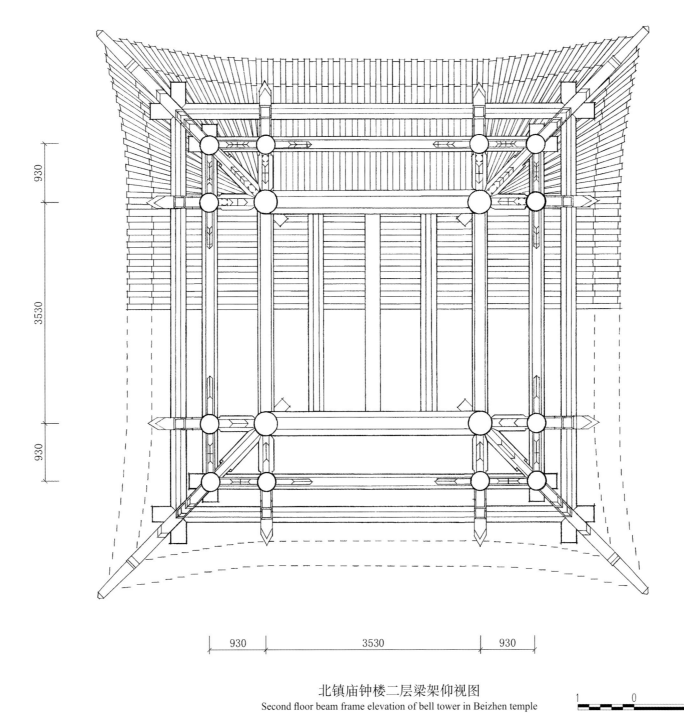

930

3530

930

930

930

3530

930

北镇庙钟楼二层梁架仰视图
Second floor beam frame elevation of bell tower in Beizhen temple

1　　0　　1　　2m

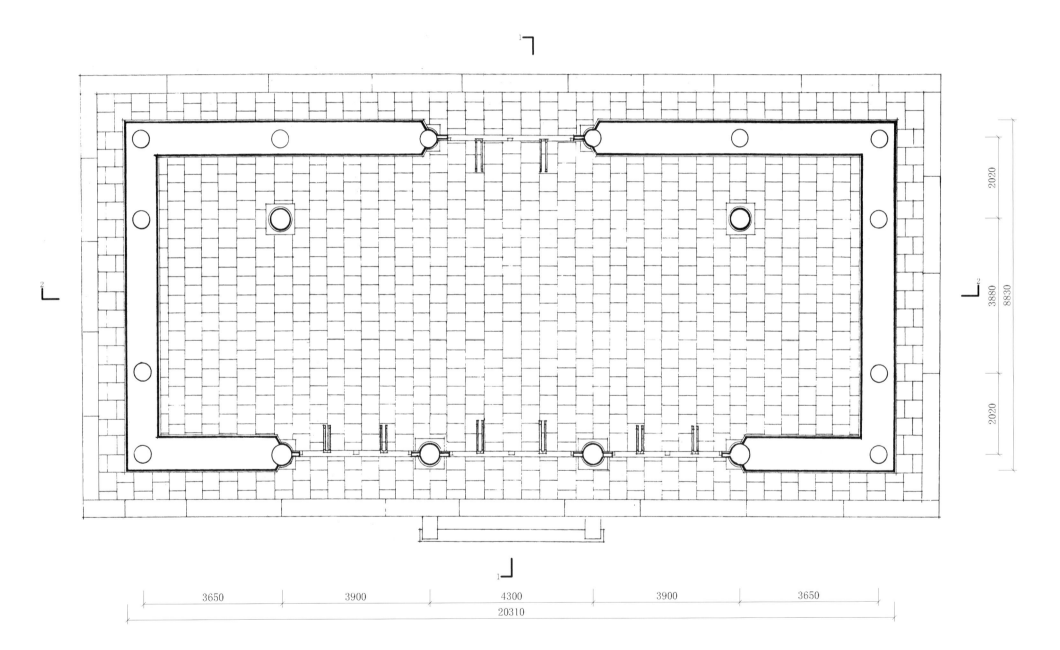

北镇庙御香殿平面图
Plan of Yuxiang hall in Beizhen temple

北镇庙御香殿正立面图
Elevation of Yuxiang hall in Beizhen temple

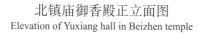

0   0.5   3m

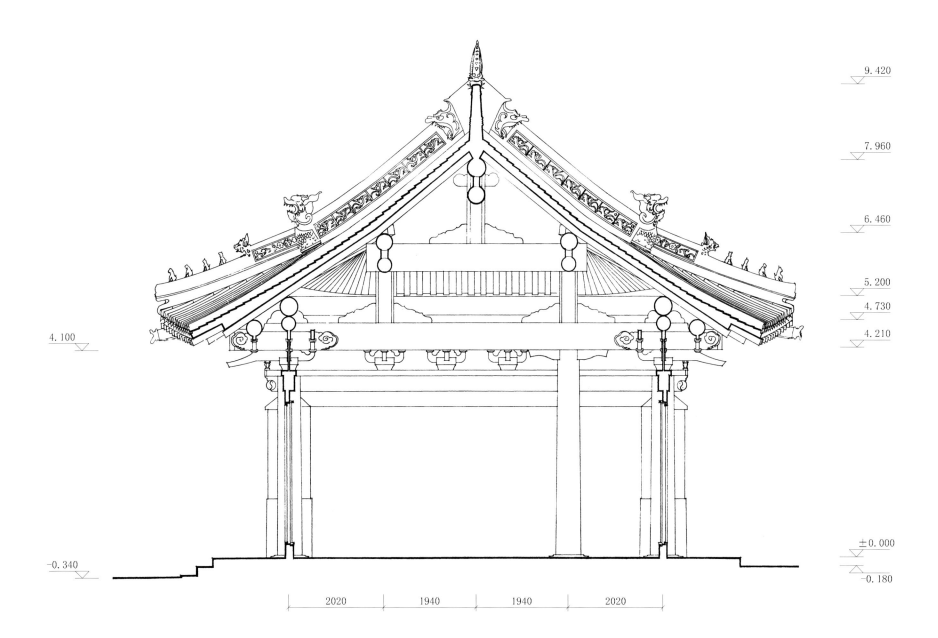

9.420

7.960

6.460

5.200

4.730

4.210

4.100

±0.000

-0.180

-0.340

| 2020 | 1940 | 1940 | 2020 |

北镇庙御香殿 1-1 剖面图

1-1 section of Yuxiang hall in Beizhen temple

北镇庙御香殿 2—2 剖面图

2-2 section of Yuxiang hall in Beizhen temple

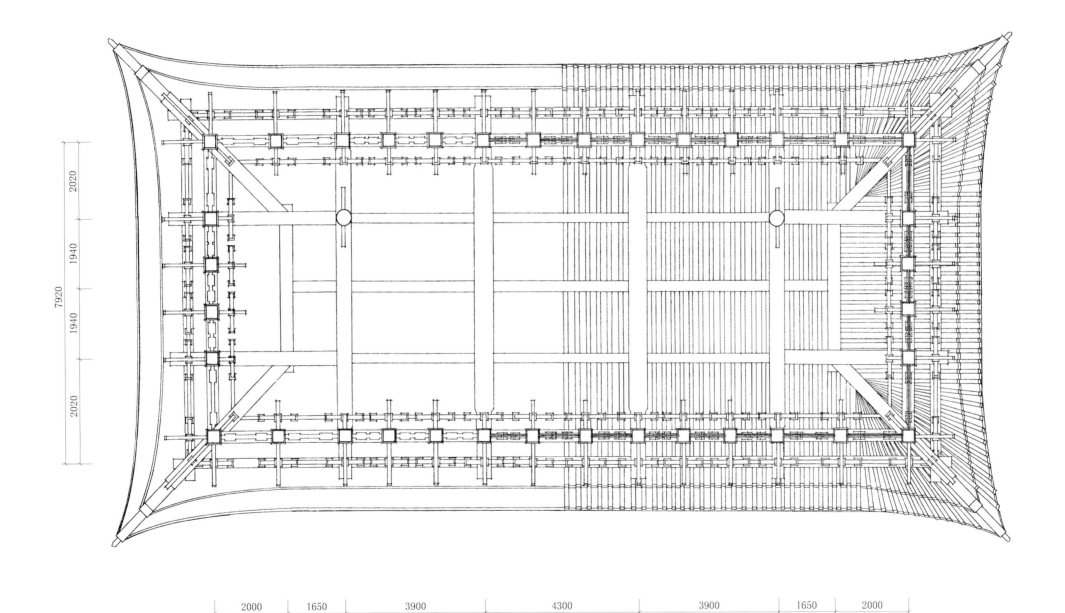

北镇庙御香殿梁架仰视图
Beam frame elevation of Yuxiang hall in Beizhen temple

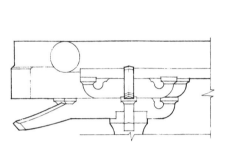
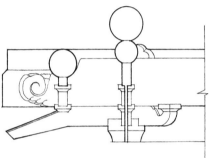
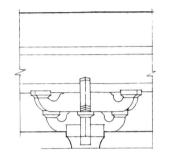
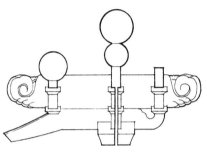
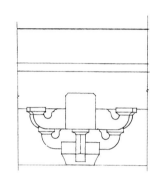
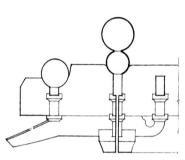

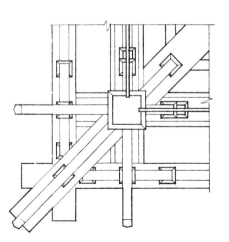

角科斗栱主要构件尺寸表 (mm)			
构件名称	高	宽	长
坐斗	250	420	420
槽升子	80	180	180
三才升	80	180	180
十八斗	80	180	180
厢棋	220	110	1030
要头带正心万棋	220	110	1870
搭角正头昂带正心瓜棋	220	110	1820
斜昂带翘	220	110	2580

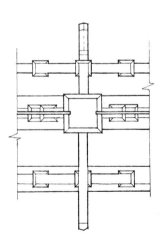

平身科斗栱主要构件尺寸表 (mm)			
构件名称	高	宽	长
坐斗	250	420	420
槽升子	80	180	180
三才升	80	180	180
十八斗	80	180	180
厢棋	220	110	1090
里拽厢棋	220	110	1090
正心瓜棋	220	110	890
正心万棋	220	110	1170
昂带翘	220	110	1840
要头	350	110	2230

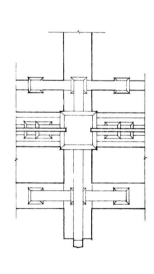

柱头科斗栱主要构件尺寸表 (mm)			
构件名称	高	宽	长
坐斗	250	420	420
槽升子	80	180	180
三才升	80	180	180
十八斗	80	180	180
厢棋	220	110	1090
里拽厢棋	220	110	1090
正心瓜棋	220	110	890
正心万棋	220	110	1170
正心枋	250	110	
里拽枋	220	110	

北镇庙御香殿斗栱大样图
Detail drawing of Dougong on Yuxiang hall in Beizhen

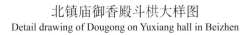

0 20   120cm

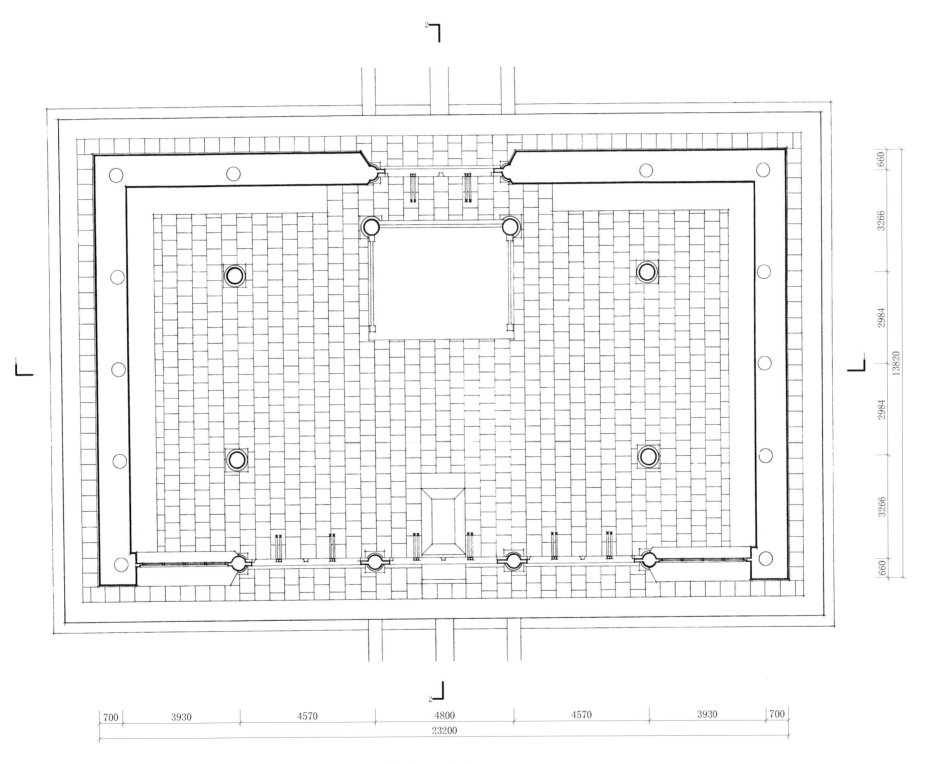

660
3266
2984
13820
2984
3266
660

700 3930 4570 4800 4570 3930 700
23200

北镇庙正殿平面图
Plan of main hall in Beizhen temple

北镇庙正殿正立面图
Elevation of main hall in Beizhen temple

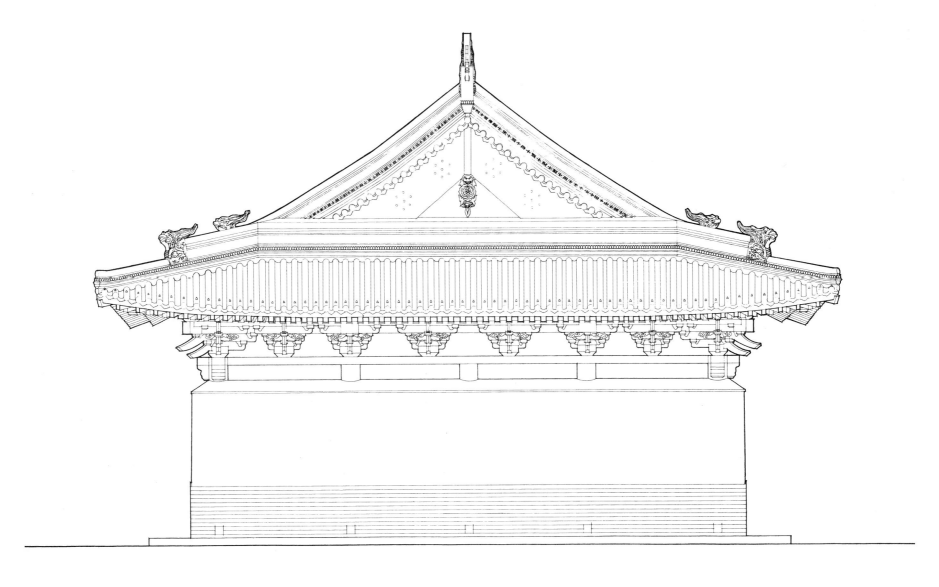

0   0.5   3m

北镇庙正殿侧立面图
Side elevation of main hall in Beizhen temple

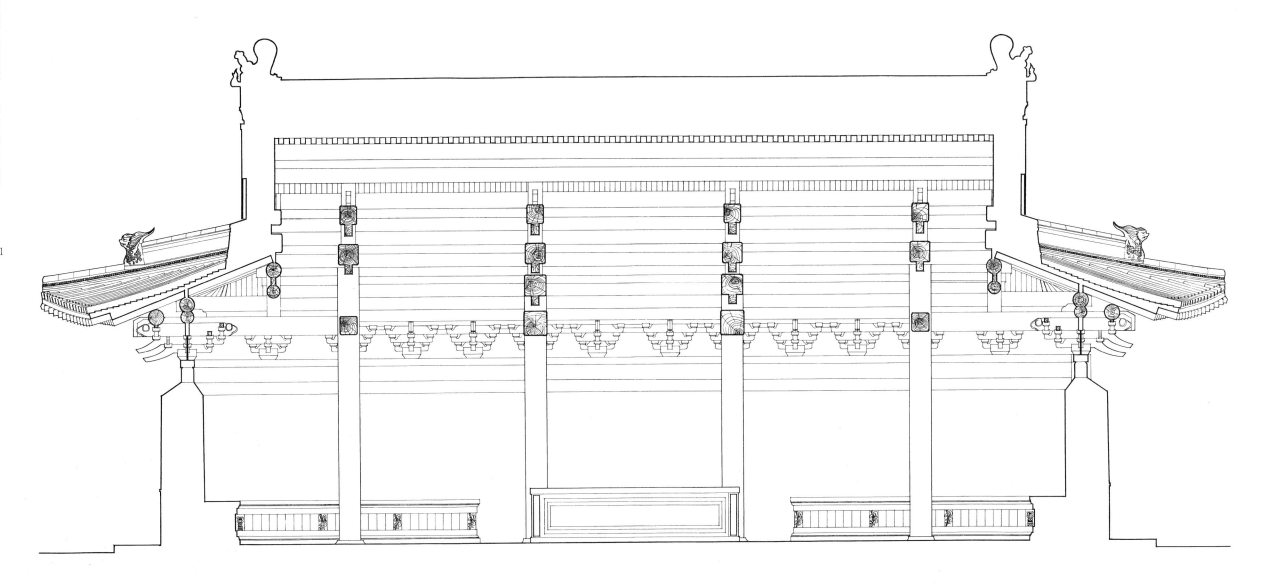

北镇庙正殿 1-1 剖面图
1-1 section of main hall in Beizhen temple

0   0.5   3m

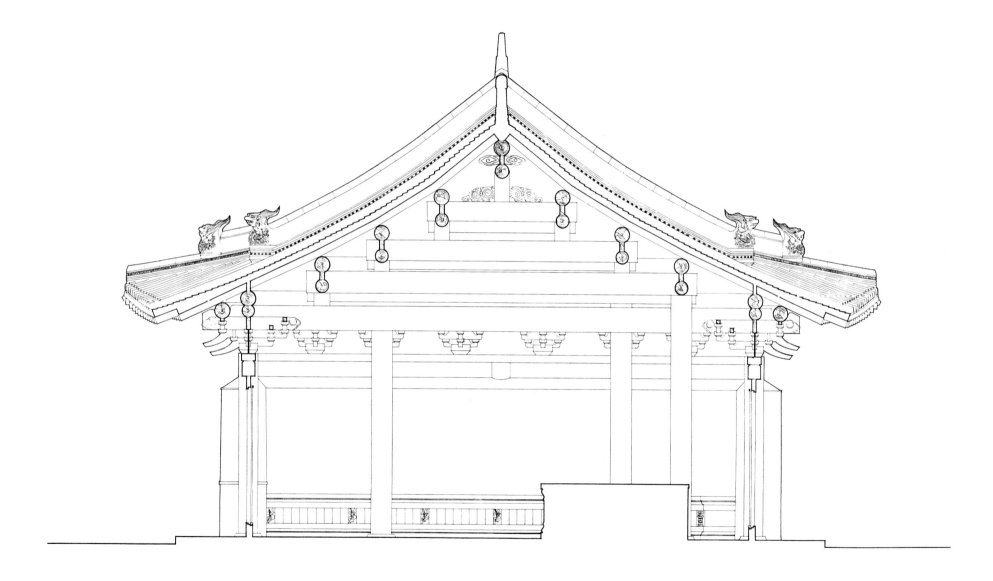

0   0.5   3m

北镇庙正殿 2-2 剖面图
2-2 section of main hall in Beizhen temple

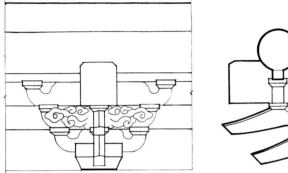
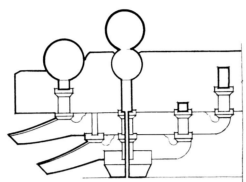
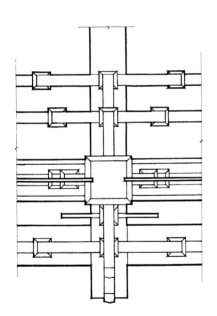

柱头科斗栱主要构件 尺寸表 (mm)			
构件名称	高	宽	长
坐斗	250	450	450
槽升子	80	180	180
三才升	80	180	180
十八斗	80	180	180
三幅云	210	110	1010
正心瓜棋	210	110	850
正心万棋	210	110	1070
厢棋	210	110	1350
单材瓜棋	210	110	1070
里拽万棋	210	110	
里拽万棋	100	110	
头昂带翘	210	110	1500
二昂带翘	210	110	2660

北镇庙正殿斗栱大样图（一）

Detail drawing of Dougong on main hall in Beizhen temple（Ⅰ）

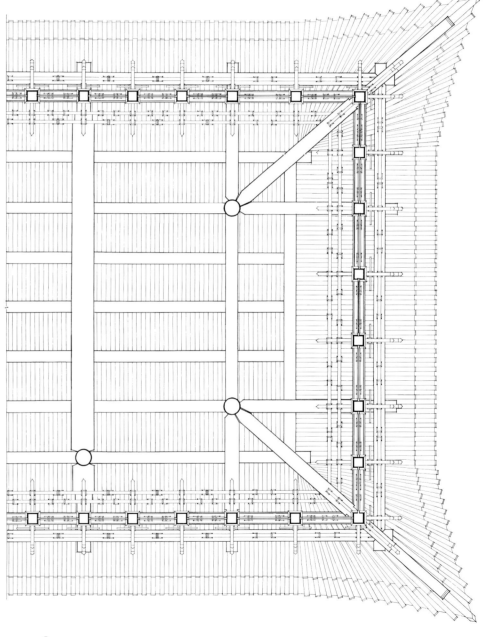

北镇庙正殿梁架仰视图

Beam frame elevation of main hall in Beizhen temple

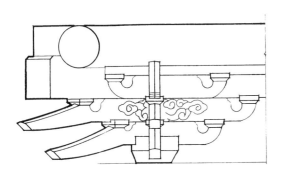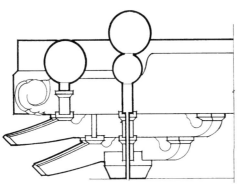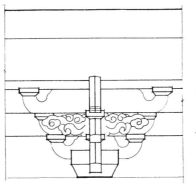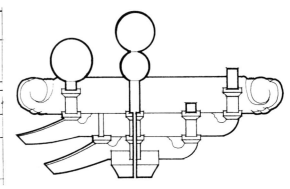

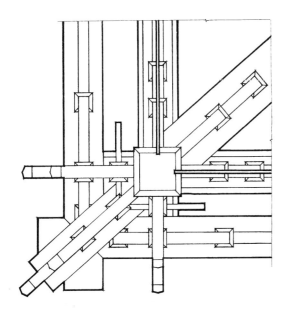

角科斗栱主要构件尺寸表（mm）			
构件名称	高	宽	长
坐斗	250	450	450
槽升子	80	180	180
三才升	80	180	180
十八斗	80	180	180
斜角头昂带翘	210	110	2150
由昂带翘	210	110	2960
搭角正头昂带正心瓜栱	210	110	1510
搭角二昂带正心万栱	210	110	2060
耍头	320	110	1030

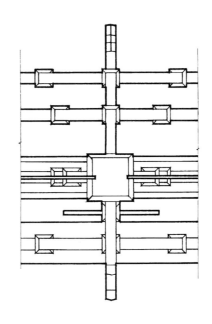

柱头科斗栱主要构件尺寸表（mm）			
构件名称	高	宽	长
坐斗	250	450	450
槽升子	80	180	180
三才升	80	180	180
十八斗	80	180	180
三幅云	210	110	1010
厢栱	210	110	1350
正心瓜栱	210	110	850
正心万栱	210	110	1070
头昂带翘	210	110	1500
二昂带翘	210	110	2060
单材瓜栱	210	110	1070
里拽厢栱	210	110	1350
耍头	320	110	2410

0   20   120cm

北镇庙正殿斗栱大样图（二）
Detail drawing of Dougong on main hall in Beizhen temple (II)

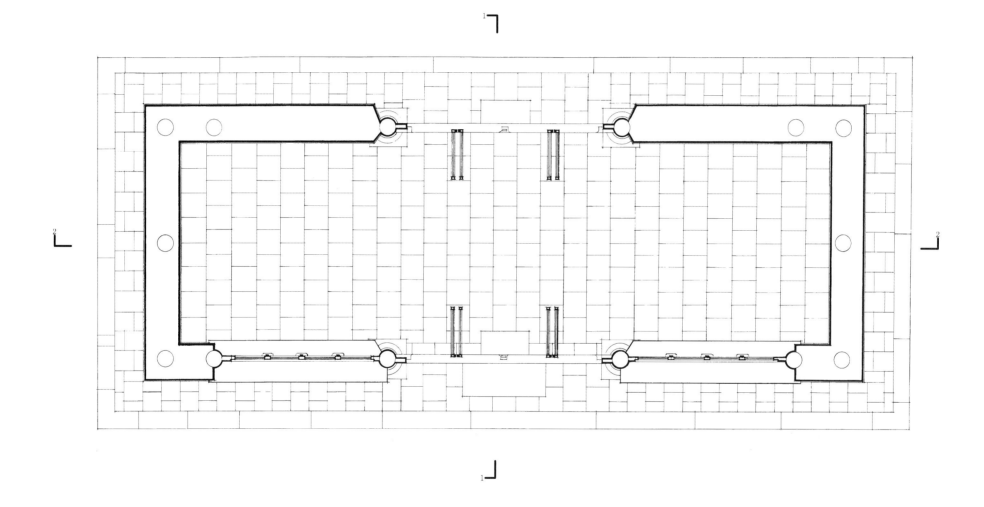

北镇庙更衣殿平面图
Plan of dressing hall in Beizhen temple

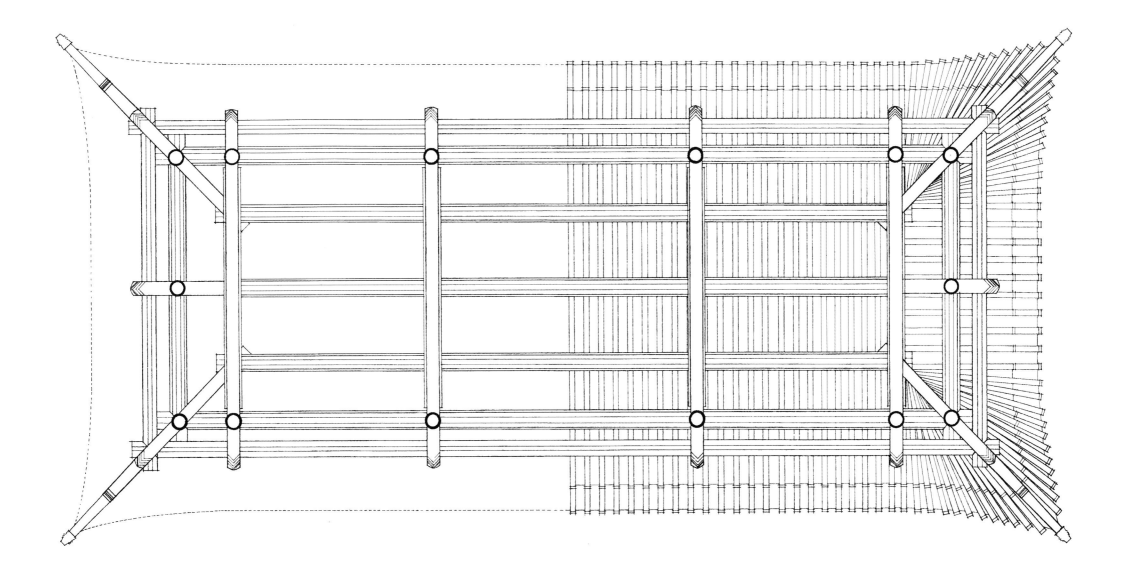

0 0.5   3m

北镇庙更衣殿梁架仰视图
Beam frame elevation of dressing hall in Beizhen temple

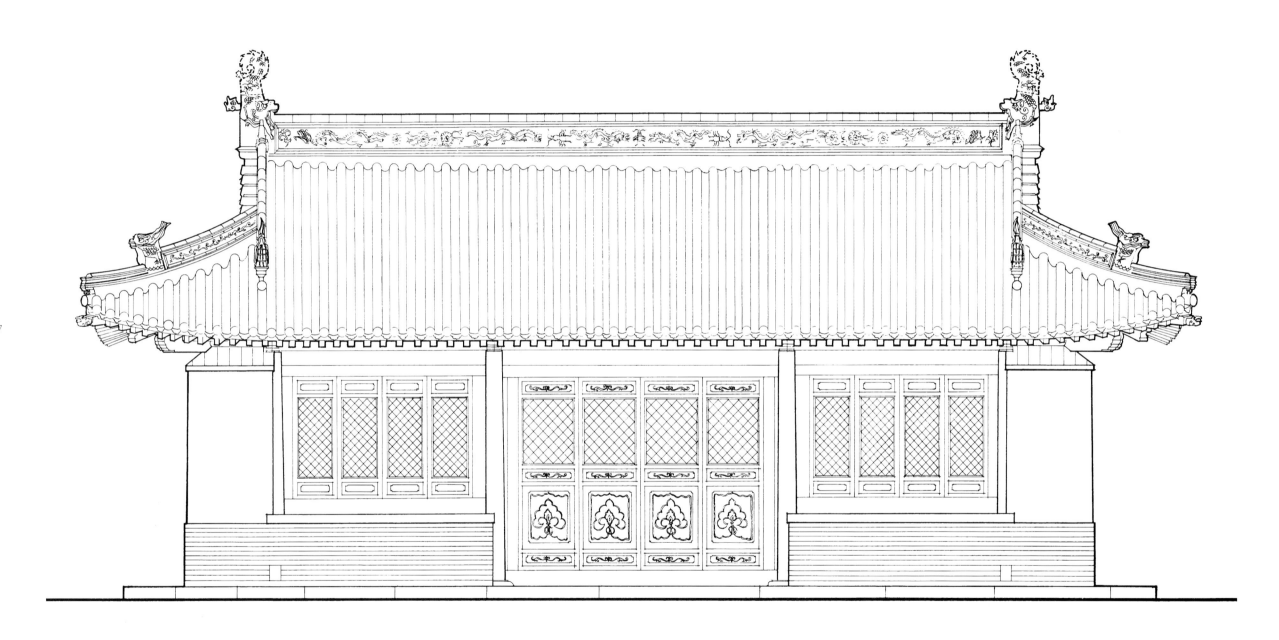

北镇庙更衣殿正立面图

Elevation of dressing hall in Beizhen temple

0   0.5   3m

0   0.5   3m

北镇庙更衣殿侧立面图
Side elevation of dressing hall in Beizhen temple

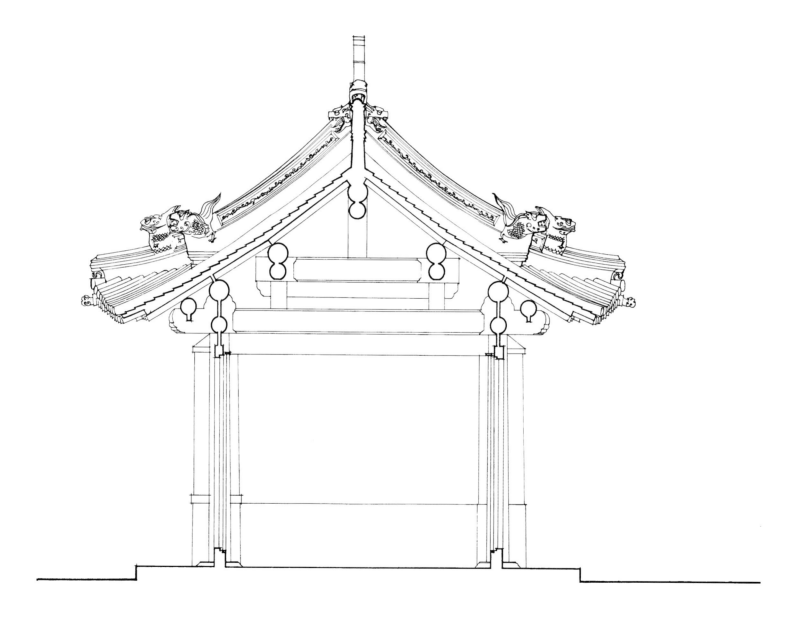

北镇庙更衣殿 1−1 剖面图

1-1 section of dressing hall in Beizhen temple

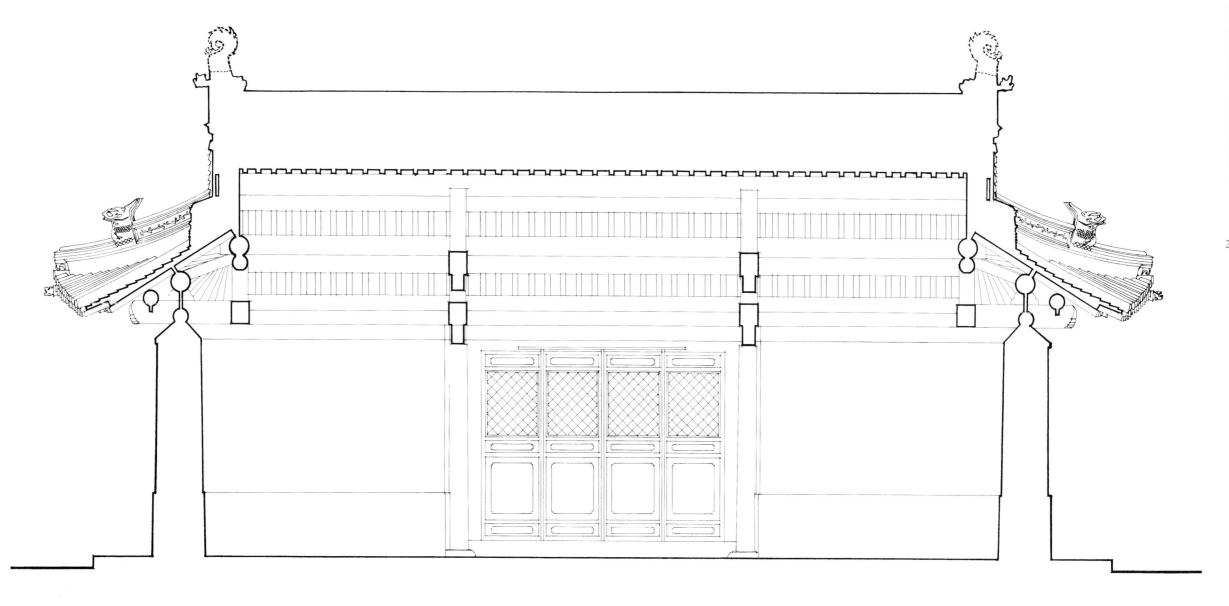

0 0.5 3m

北镇庙更衣殿 2-2 剖面图

2-2 section of dressing hall in Beizhen temple

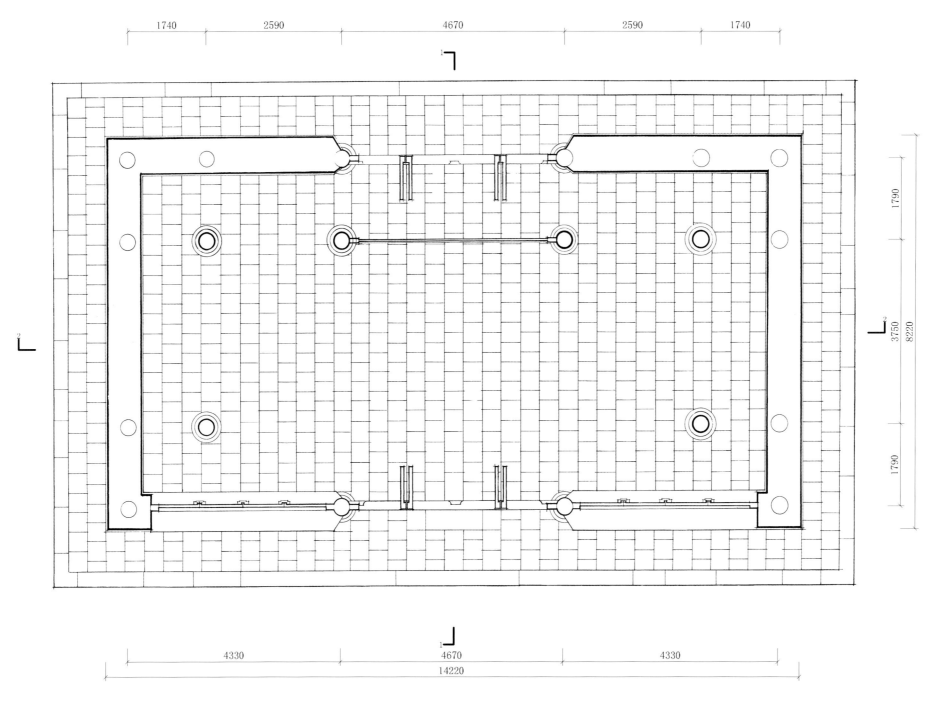

北镇庙内香殿平面图
Plan of Neixiang hall in Beizhen temple

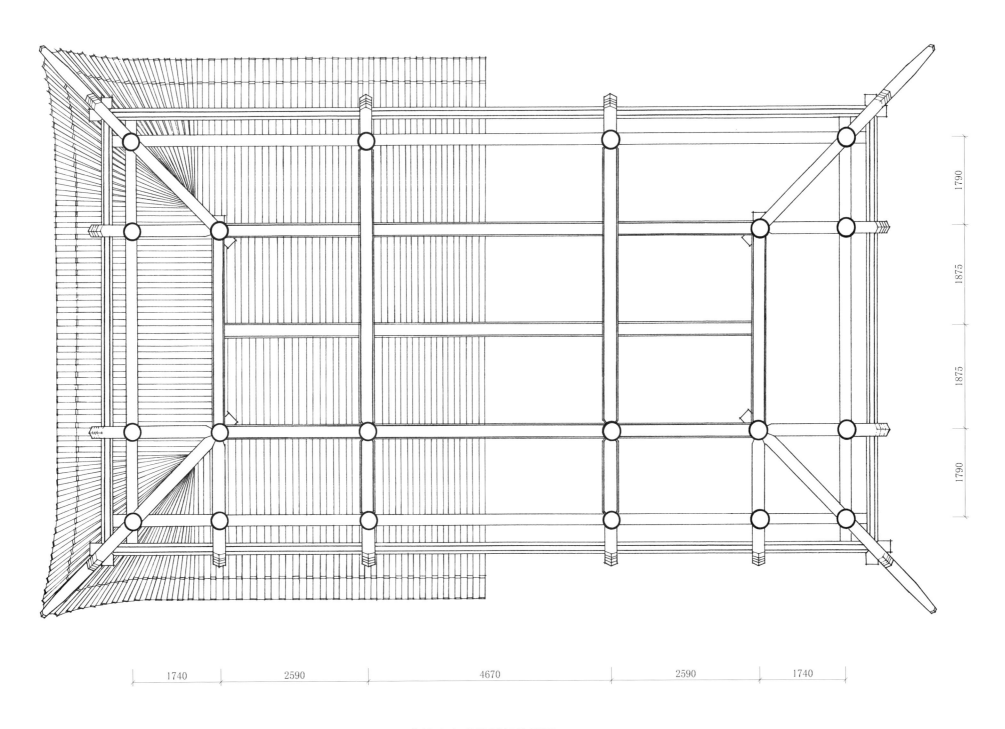

1790

1875

1875

1790

1740　2590　4670　2590　1740

北镇庙内香殿梁架仰视图

Beam frame elevation of Neixiang hall in Beizhen temple

北镇庙内香殿正立面图

Elevation of Neixiang hall in Beizhen temple

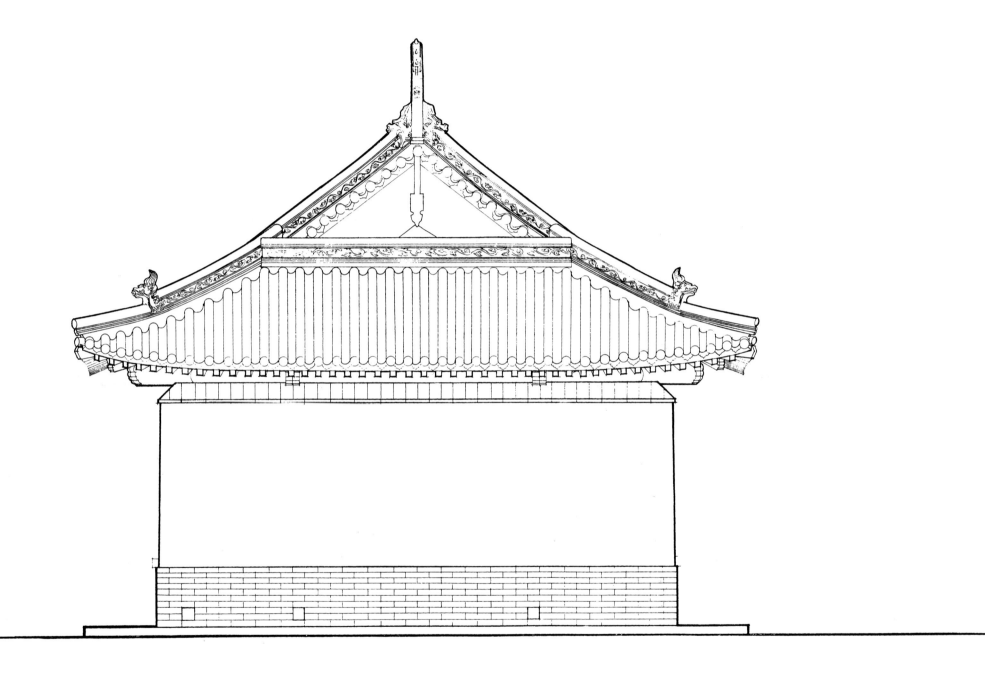

0   0.5   3m

北镇庙内香殿侧立面图
Side elevation of Neixiang hall in Beizhen temple

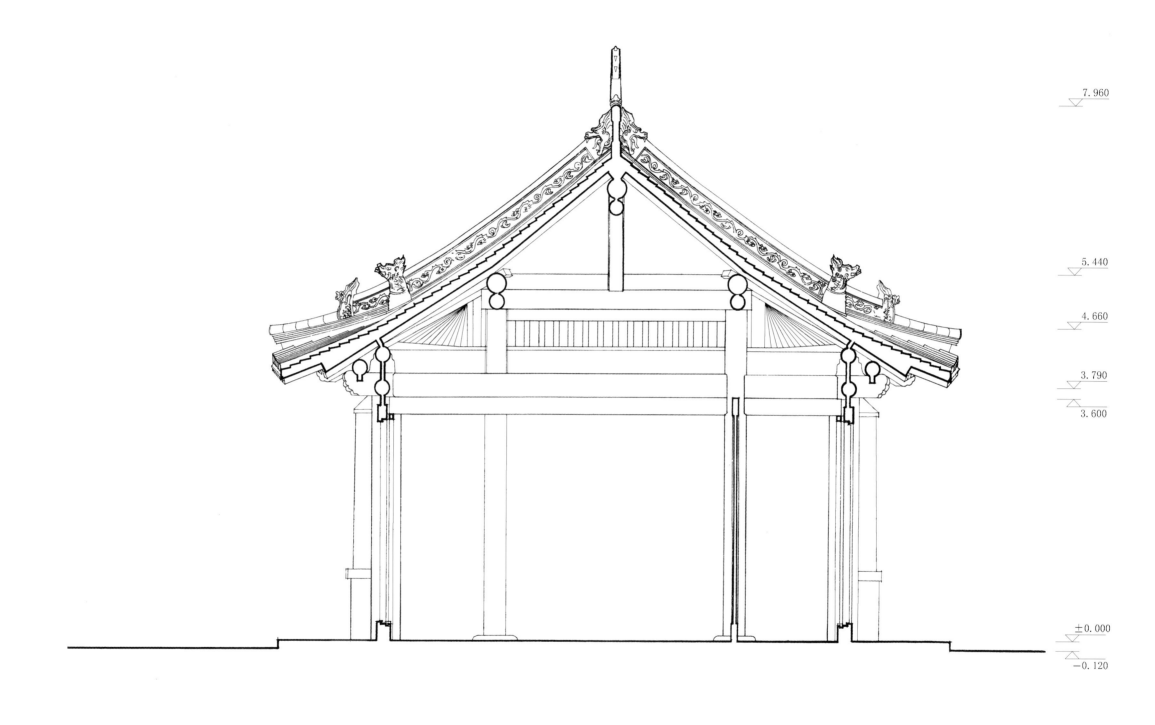

7.960

5.440

4.660

3.790

3.600

±0.000

−0.120

北镇庙内香殿 1−1 剖面图
1-1 section of Neixiang hall in Beizhen temple

0   0.5   3m

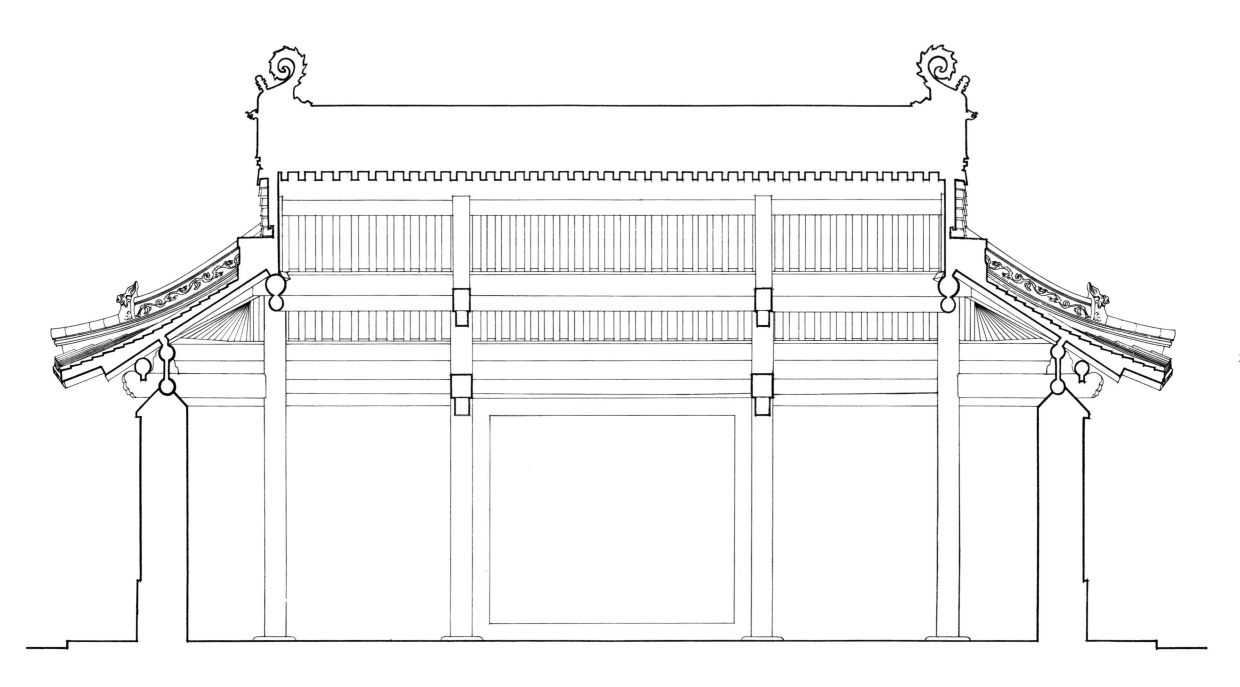

0   0.5   3m

北镇庙内香殿 2−2 剖面图
2-2 section of Neixiang hall in Beizhen temple

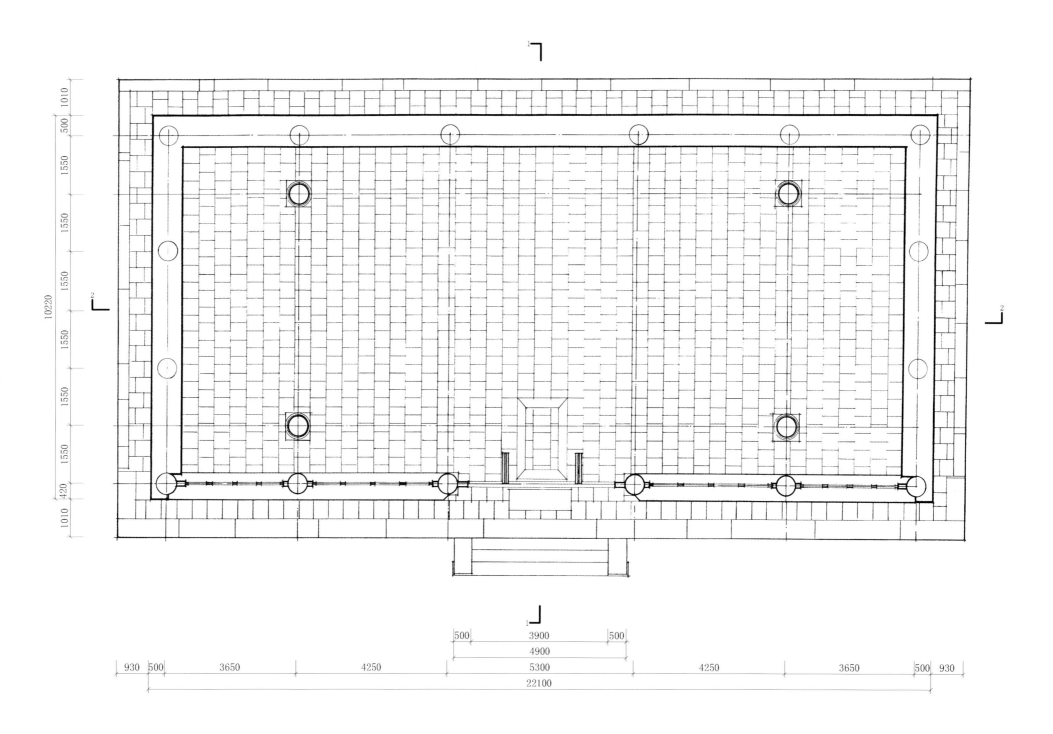

北镇庙寝殿平面图
Plan of resting hall in Beizhen temple

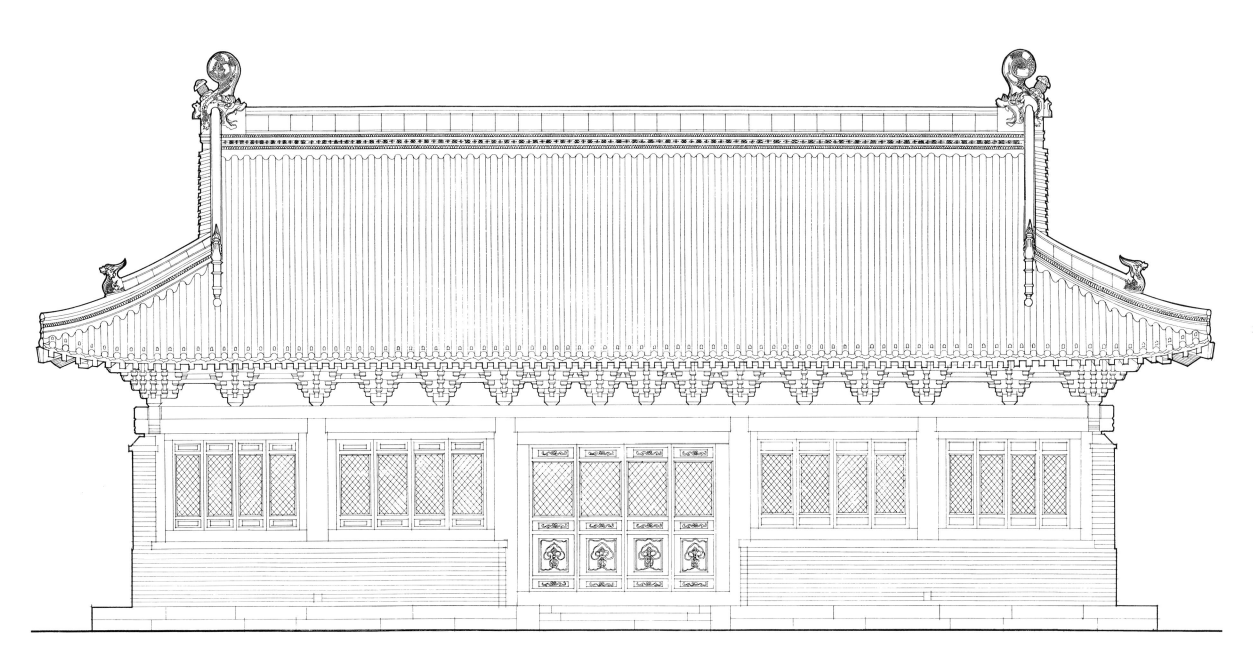

北镇庙寝殿正立面图
Elevation of resting hall in Beizhen temple

北镇庙寝殿侧立面图
Side elevation of resting hall in Beizhen temple

0   0.5   3m

北镇庙寝殿 1-1 剖面图
1-1 section of resting hall in Beizhen temple

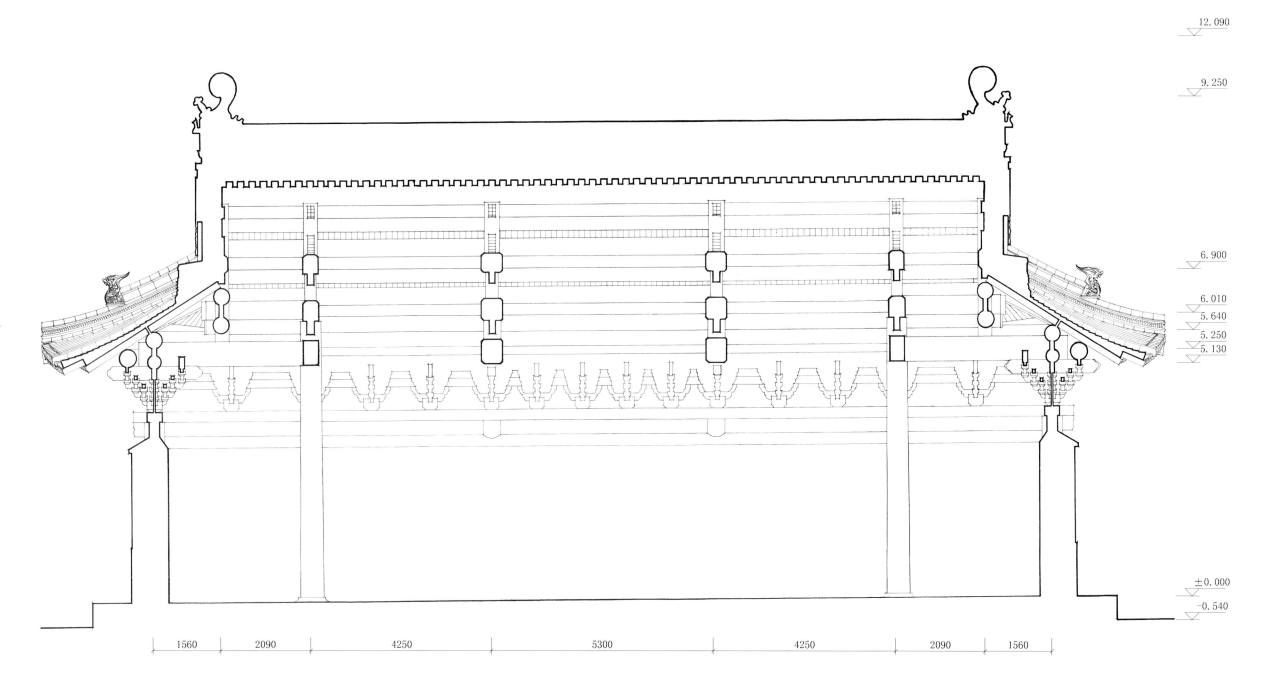

12.090

9.250

6.900

6.010
5.640
5.250
5.130

±0.000

−0.540

1560 2090 4250 5300 4250 2090 1560

北镇庙寝殿 2−2 剖面图

2-2 section of resting hall in Beizhen temple

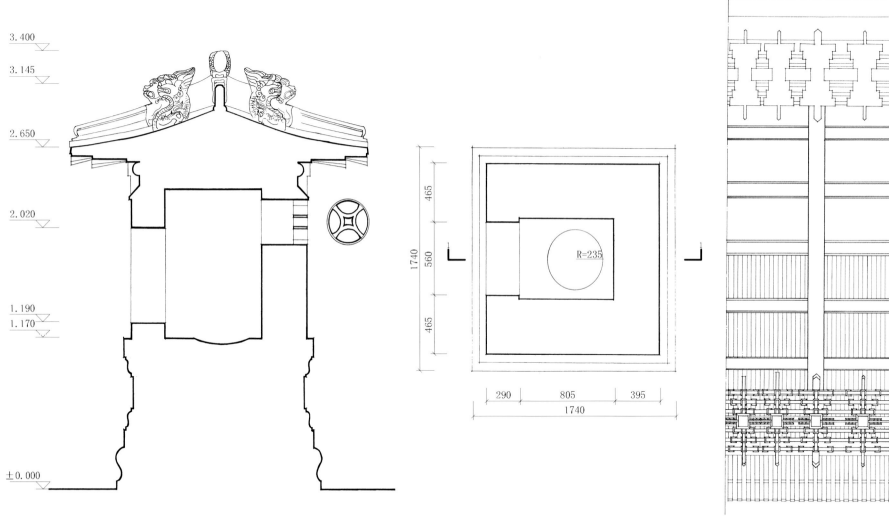

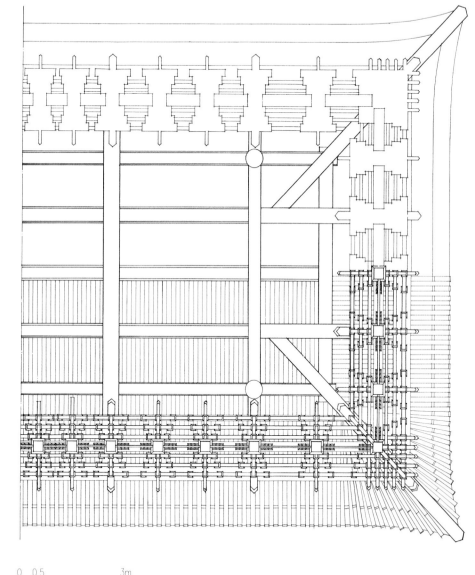

北镇庙焚帛炉 1−1 剖面图

1-1 section of stove for burning in Beizhen temple

北镇庙焚帛炉平面图

Plan of stove for burning in Beizhen temple

北镇庙寝殿梁架仰视图

Beam frame elevation of resting hall in Beizhen temple

北镇庙焚帛炉正立面图
Elevation of stove for burning in Beizhen temple

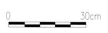

0   30cm

北镇庙焚帛炉侧立面图
Side elevation of stove for burning in Beizhen temple

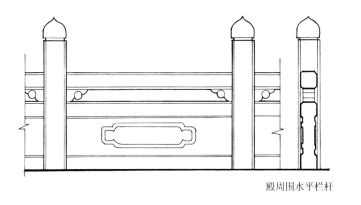

殿周围水平栏杆

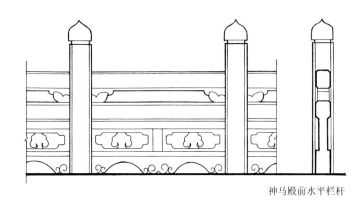

神马殿前水平栏杆

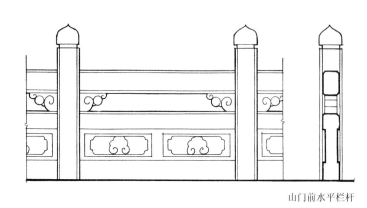

山门前水平栏杆

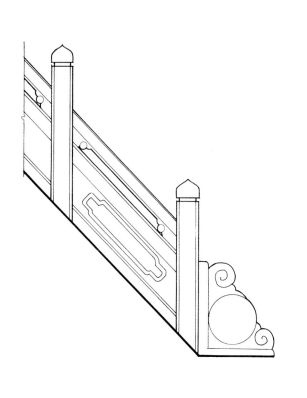

殿周围栏板及抱鼓石

御香殿前抱鼓石

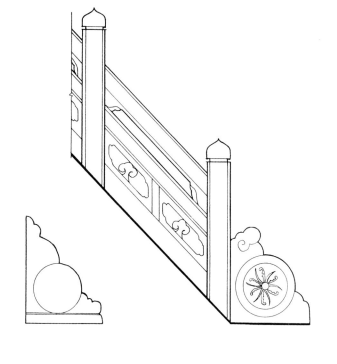

神马殿前斜栏杆及抱鼓石

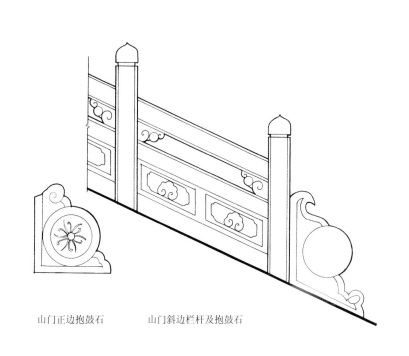

山门正边抱鼓石　　　山门斜边栏杆及抱鼓石

北镇庙栏杆大样图
Detail drawings of rails in Beizhen temple

0 0.2　　　　　　1m

参与测绘及相关工作的人员名单

一、1981 年清永陵测绘

指导教师：侯幼彬　陶友松　曹汛

参与测绘（1979 级学生）：

李江宏　周琳　朱小东　任卫东　任炳文　甄兰平　张军燕

袁安江　赵娟　刘丽娜　周艺玲　李华哲　吴向群

王琦　汤卓如　武小勇　吴黎　郭保宁　杨光　卢祥　刘战

吴岩松　田大方　田原　付曼丽　赵永丰　王洪达　杨积洪　田铁军

姜长斌（进修生）　王践诺（进修生）　迟培成（进修生）

二、1985 年清福陵测绘

指导教师：侯幼彬　陶友松　曹汛　刘大平（硕士生）　邹广天（硕士生）

参与测绘（1982 级学生）：

陈列殷　李健红　宋谦　倪娥　夏为

崔永祥　冯珊　李英　巩智民　姜波　关宏明

三、1984 年清昭陵测绘

指导教师：侯幼彬　陶友松　曹汛　刘大平（硕士生）　邹广天（硕士生）

参与测绘（1981 级学生）：

王志民　王玮　李玲玲　邵龙　白小鹏　杨雪飞

李为　姚谦　王崇民　李笑岩　郝之颖　段旺　刘嘉华　艾杰

曾钰　唐克铮　宋晓龙　褚英　侯其明　王砺　丁立民　张安可

周松华　张春阳　孙一民　王之光　谭军　张弛　蒋群力

许伟　顾倩予　张健　魏强　王洪义　王丽香　周福杰　黄更新

王奎仁　于学军　刘昕　田景　郭景　王叶民

四、1981 年北镇庙测绘

指导教师：侯幼彬　陶友松　曹汛　赵光辉（硕士生）

参与测绘（1978 级学生）：

马斌　刘滨　陈原宏

陈邦贤　陈昌生　范雪　高薇　葛蔚平　刘大平

五、图纸整理及翻译人员

刘劲松　李曼曼　陆　伟　李晓霁　梅洪元　孟　磊　任　明　沈黛琦
唐东宁　王宏新　王　君　王少梅　王晓光　吴雅娟　王正刚　徐苏宁
谢忠辉　周　畅　周大荣　邹广天　张姗姗　郑晓洪　张晓玲

图纸整理：李　琦　曲　蒙　司道光　何璐西　张书铭　郭　威　赵庆超　宗　敏
　　　　　周　楠　息　琦　陈　晨　陈一鸣　刘方溪　朱彦涵　张成磊　朱　萌
　　　　　于政委　徐见卓　陈永臻　张宁戈　胡一鸣　夏月亮　孙　岩　房晨璇
　　　　　周　媛

翻译统筹：刘　洋　[奥]荷雅丽

英文翻译：王　岩　王秋玉　何璐西　刘文卿　[奥]荷雅丽　Michael Norton

Name List of Participants Involved in Surveying and Related Works

Yong Mausoleum Surveying and Mapping in 1981

Supervising Instructor: HOU Youbin, TAO Yousong, CAO Xun, ZHAO Guanghui(Postgraduate student)

Team Members (Students of 1979): YUAN Anjiang, ZHAO Juan, LIU Li'na, ZHOU Yiling, LI Huazhe, WU Xiangqun, LI Jianghong, ZHOU Lin, ZHU Xiaodong, REN Weidong, HAN Yahui, REN Bingwen, ZHEN Lanping, ZHANG Junyan, WANG Qi, TANG Zhuoru, WU Xiaoyong, WU Li, GUO Baoning, YANG Guang, LU Xiang, LIU Zhan, WU Yansong, TIAN Dafang, TIAN Yuan, FU Manli, ZHAO Yongfeng, WANG Hongda, YANG Jihong, TIAN Tiejun, JIANG Changbin(Visiting student), WANG Jiannuo(Visiting student), CHI Peicheng(Visiting student)

Fu Mausoleum Surveying and Mapping in 1985

Supervising Instructor: HOU Youbin, TAO Yousong, CAO Xun, LIU Daping(Postgraduate student), ZOU Guangtian(Postgraduate student)

Team Members (Students of 1982): CHEN Lie, YIN Xin, LI Jianhong, SONG Qian, NI E, XIA Wei, CUI Yongxiang, FENG Shan, LI Ying, GONG Zhimin, JIANG Bo, GUAN Hongmin

Zhao Mausoleum Surveying and Mapping in 1984

Supervising Instructor: HOU Youbin, TAO Yousong, CAO Xun, LIU Daping(Postgraduate student), ZOU Guangtian(Postgraduate student)

Team Members (Students of 1981): WANG Zhimin, WANG Wei, LI Lingling, SHAO Long, BAI Xiaopeng, YANG Xuefei, LI Wei, YAO Qian, WANG Chongmin, LI Xiaoyan, HAO Zhiying, DUAN Wang, LIU Jiahua, AI Jie, ZENG Yu, TANG Kezheng, SONG Xiaolong, CHU Ying, HOU Qiming, WANG Li, DING Limin, ZHANG Anke, ZHOU Songhua, ZHANG Chunyang, SUN Yimin, WANG Zhiguang, TAN Jun, ZHANG Chi, JIANG Qunli

Team Members (Students of 1982): WANG Kuiren, YU Xuejun, LIU Xin, TIAN Jian, GUO Jing, WANG Yemin, XU Wei, GU Qianyu, ZHANG Jian, WEI Qiang, WANG Hongyi, WANG Lixiang, ZHOU Fujie, HANG Gengxin, MA Bin, LIU Bin, CHEN Yuanhong

Beizhen Temple Surveying and Mapping in 1981

Supervising Instructor: HOU Youbin, TAO Yousong, CAO Xun, ZHAO Guanghui(Postgraduate student)

Team Members (Students of 1978): CHEN Bangxian, CHEN Changsheng, FAN Xue, GAO Wei, GE Weping, LIU Daping, LIU Jinsong, LI Manman, LU Wei, LI Xiaoji, MEI Hongyuan, MENG Lei, REN Ming, SHEN Daiqi, TANG Dongning, WANG Hongxin, WANG Jun, WANG Shaomei, WANG Xiaoguang, WU Yajuan, WANG Zhenggang, Xu Suning, XIE Zhonghui, ZHOU Chang, ZHOU Darong, ZOU Guangtian, ZHANG Shanshan, ZHENG Xiaohong, ZHANG Xiaoling

Editor of Drawings and Translation

Editor of Drawings: LI Qi, QU Meng, SI Daoguang, HE Luxi, ZHANG Shuming, GUO Wei, ZHAO Qingchao, ZONG Min, ZHOU Nan, XI Qi, CHEN Chen, CHEN Yiming, LIU Fangxi, ZHU Yanhan, ZHANG Chenglei, ZHU Meng, YU Zhengwei, XU Jianzhuo, CHEN Yongzhen, ZHANG Ningge, HU Yiming, XIA Yueliang, SUN Yan, FANG Chenxuan, ZHOU Yuan

Translator in Chief: LIU Yang, Alexandra Harrer

Team Members: WANG Yan, WANG Qiuyu, HE Luxi, LIU Wenqing, Alexandra Harrer, Michael Norton

图书在版编目(CIP)数据

关外三陵和北镇庙=THREE MAUSOLEUMS OUTSIDE
SHANHAIGUAN PASS AND BEIZHEN TEMPLE/侯幼彬，刘大
平，刘洋主编；哈尔滨工业大学建筑学院编写．—北京：
中国建筑工业出版社，2018.6
　　（中国古建筑测绘大系．陵寝建筑与祠庙建筑）
　　ISBN 978-7-112-22281-0

　　I.①关… II.①侯… ②刘… ③刘… ④哈… III.
①陵墓-建筑艺术-辽宁-清代-图集 IV.
①TU251.2-64

中国版本图书馆CIP数据核字（2018）第114775号

丛书策划／王莉慧
责任编辑／李　鸽　陈海娇
英文审稿／［奥］荷雅丽（Alexandra Harrer）
书籍设计／付金红
责任校对／王　烨

中国古建筑测绘大系·陵寝建筑与祠庙建筑

关外三陵和北镇庙

哈尔滨工业大学建筑学院　编写
侯幼彬　刘大平　刘　洋　主编

Traditional Chinese Architecture Surveying and Mapping Series:
Mausoleum Architecture & Shrines and Temples Architecture
THREE MAUSOLEUMS OUTSIDE SHANHAIGUAN PASS AND BEIZHEN TEMPLE
Compiled by School of Architecture, Harbin Institute of Technology
Edited by HOU Youbin, LIU Daping, LIU Yang

＊

中国建筑工业出版社出版、发行（北京海淀三里河路9号）
各地新华书店、建筑书店经销
北京方舟正佳图文设计有限公司制版
北京雅昌艺术印刷有限公司印刷

＊

开本：787毫米×1092毫米　横1／8　印张：31　字数：821千字
2020年5月第一版　2020年5月第一次印刷
定价：238.00元
ISBN 978-7-112-22281-0
　　　　（31663）